Life in the Lab
The Essential Guide to Becoming a Researcher

Experimental design, conducting research, data analysis and statistics, and writing up your work for publication

First Edition

Kevin J Hamill

Any resemblance to actual persons living or dead, events, or locales is entirely coincidental.

For further information and to contact the author visit:
www.lantsandlaminins.com
or email:
lantsandlaminins@gmail.com

Preface to Edition 1.0

I adopted a non-traditional publishing method for this book, releasing chapters semi-regularly during 2019 and 2020 via leanpub.com. The feedback from early readers shaped the direction of the book and during lockdown of 2020 I updated everything. Of course, after that there was another round of edits in 2022. As with all writing, it is hard to let go and hit "submit"! There *will* always be more editing to do. Therefore, if you have comments or suggestions for how the content could be improved or expanded then please do get in touch.

Acknowledgements

The book would not have happened if it hadn't been for invaluable support and input from all the people who read and commented on early editions. Special acknowledgements go to Susan Hamill, Hilary Currin, Dr Liam Shaw, Dr Lee Troughton, Conor Sugden, Olivia Kingston, Luke Tasker-Lynch, Dr Ben Boroski and Dr Joseph Michael. This team of stalwarts helped shape the content of the book more than anyone else. They also spotted many of the errors. Any mistakes remaining in this version are mine. However, to be fair, I did find a passage in the latest edit where my cat contributed without my approval. Don't worry, it has been removed.

Table of Contents

Chapter 1: Before you Begin

1.1 About this book
Description of the book contents and how to make the most of the material within it.

1.2 Choosing a Research Project
What you should consider when deciding which lab to work in and what project to pursue.

1.3 Applying for a Research Position
Advice for preparing job applications, including some interview questions you should expect.

1.4 Starting a new project
Quick tips for when you are brand new to a lab

1.5 Information gathering
Identifying the resources you will need, where to find them, and how to assess their quality.

1.1 About this book

My goal for this book was to create a useful resource for people embarking on a research career. The "one book that every research student should read". I have included all the different aspects of life as a researcher starting from before you begin right through to publishing high-quality papers. The content is deep enough to be useful but should be accessible and entertaining enough to not be boring or difficult to read.

The contents of this book are intended to lay solid foundations for all the main aspects of being a successful scientist. Your supervisors, lab mates, mentors and lecturers will all build on the foundations. You will also need to go deeper into your project-specific areas. Hopefully, by reading this book, everything else should be easier and quicker to learn. If I have done a good job, those tricky statistics lessons, experimental design workshop or complicated textbooks will all be less painful.

How to use this book

The first few chapters, about life in the lab and experimental design, are so fundamentally essential to becoming a good researcher that I recommend reading these parts in the order they are presented.

The later chapters go into more of the specific details on how to analyse data, how to prepare figures, how to write about and present your work. Again, these parts *can* be read cover-to-cover, but each is also entirely capable of standing alone. Therefore, I recommend that whenever you have something specific to write or prepare, dip into the relevant section immediately before doing the work to refresh your knowledge on what to do. I have tried to keep each section short but as comprehensive as possible, and each part should be quick and easy to read.

Mostly

Throughout, I have tried to keep the information as general as possible, but please remember that we work in a large discipline and there will almost always be exceptions to any specific point being raised.

Writing, in particular, is an area where the "rules" are really just guidelines. You should always consider the advice given here in light of your own situation and then do what is best for you or your data. Everything in the book is fundamentally good advice, but some of it may only "mostly" refer to your specific situation!

About the author

I am a Senior Lecturer at the University of Liverpool in the UK, where I teach "Laboratory Skills" to Masters students, "Experimental Design" to PhD students, and where I run a variety of writing courses for postgraduate and undergraduate students. I currently have five PhD students working in my laboratory and, each year, I also supervise three to four undergraduate and four to five Masters-level project students.

Much of the content in this book started life as the handouts and lecture notes from the courses I teach. As the years passed and the material grew, I realised that all these resources would be most useful gathered together into a single volume book that could be easily shared. Hence, this book.

In addition to teaching, I run a molecular biology research lab focused on cell to extracellular matrix interactions (LaNts and laminins). My team's work encompasses a range of scales from whole organism and behavioural studies, studies using human disease samples or transgenic animals, studies using ex vivo, 2D and 3D cell culture models, and on downward in scale to molecular or protein biochemistry studies including measuring alternative splicing and protein-protein interactions. I have written and reviewed many papers and serve on the editorial board on research journals within my field. Some of the examples within this book come from my research; however, the core messages apply equally well across the entire biological sciences and medical disciplines and to wider research practices in any discipline.

Before moving to the University of Liverpool, I was a post-doctoral researcher and then Research Assistant Professor at Northwestern University, Chicago. Prior to that, I obtained my PhD in Molecular Genetics from the University of Dundee, Scotland. For most things in science, the location of the lab does not make a difference. However, in this book I have attempted to point out local differences wherever these are relevant.

About the cartoons

When lecturing and producing the hand-outs for my writing courses, I quickly realised that something was needed to make the subject a little less dry. Adding cartoons seemed to help. I've based the appearance of the characters in this book on real people who have worked in my lab or my Department Professors (Profs), but the events depicted contain a lot of creative licence!

The "big tips" boxes reflect real advice from faculty members at my Institution.

The Students

The Profs

1.2   Choosing a Project

Every step along your career path involves deciding upon a course of action. These decisions can feel big and scary as the choices you make affect your career trajectory in the long term and your lifestyle in the short term. Eeek! However, these decisions are also exciting too. What's right for someone else isn't necessarily the best for you, therefore there is not one piece of advice that can work for everyone. Instead, this section highlights the things that are worth taking into consideration so that you can be confident that you make the best decision *for you*.

Project Choice

The first decision is what to work on. Here you should consider what you care about, what the project will add to the world, how well suited you are to the project, and the development opportunities offered. You should not only be making your choice based solely on what you *can do* but rather, what is *best for you*. Here are some things to consider:

Choose something you care about (that you think is worthwhile)

This advice sounds obvious, and it should be no surprise, but if you don't care about your project, then you will find it harder to stay motivated during the difficult times. The longer the project, the more critical this becomes. If you are choosing a PhD project, then caring about the work should be high up your list in terms of relative importance. However, if you are selecting a ten-week lab placement or a summer internship, then the subject of the project is not so crucial; better to focus your attention on the transferrable skills and techniques that you will learn.

But where does the "passion" come from? It really is a personal thing but here are a few topics to consider:

Discovery or Translational?

Projects might be labelled in different ways. "Discovery science", "science-led" or "basic science" usually mean that the work is directed at a more fundamental deepening our understanding of something. Note that "basic" does not mean that the approaches are simple and there is a clear move away from using this term due to these potential negative connotations. In contrast, "translational research" usually means that there is direct path to the end application. The underpinning science will have already been largely already done and now is the time to exploit that core knowledge for practical applications. Note that there is often overlap; most "discovery science" projects will have an ultimate aim of being exploited for some real-world value, but likely that onward application will be far into the future, beyond the lifetime of the project. In contrast, translational projects will also involve making discoveries and overcoming obstacles along the way but may be "objective" rather than "hypothesis" led.

If you are motivated by the thrill of the finding something new, then likely you are best to look for a "discovery science" project, or at least one with strong discovery elements. These projects are likely to be about putting together the clues from various pieces of data to make sense of it all. You might struggle to wow someone in the pub by talking about the specifics of your work, but you have a chance to cause a much more fundamental change to the way that we think about the world. Good discovery science projects provide valuable knowledge irrespective of the outcome. Important character traits are an ability to think laterally and to extrapolate from incomplete information, and a general willingness to be open to the data and "follow the science".

In contrast, you might find a project where the outcomes are more concrete and direct is more appealing. It could be that you have a personal connection to a disease or inherited condition, and so the opportunity to develop an improved treatment provides direct motivation. Similarly, improving food security, yield, land use, or protection of endangered species all have clear goals where you could ultimately *make a difference*. These sorts of projects need to give a specific outcome to be valuable…do you have a mindset that thrives on problem-solving and spotting a route to a solution?

Understandable?

I recommend choosing a project where you are not only interested but also are able to absorb the core concepts relatively easily. Things get more difficult as you go deeper! There is a big caveat here, who and how you were taught has a large impact on how you feel about a subject.

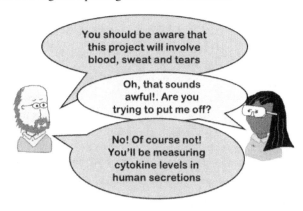

Everyone has to work hard at their studies at some point, so don't rule out anything that you are really interested in. Instead aim to strike a balance between what you will be willing to work hard for and how difficult it will be.

Fashionable?

Science, just like everything else, goes through cycles of specific topic areas being more or less fashionable. What is "hot" at any time is driven by a mixture of what large discoveries have been made lately, what new technologies have become available, and the will of

Big Tip

Just because a project Is in a "hot" area, doesn't mean that it is the right choice for you. Choose based on your needs.

the political parties or funding bodies that are controlling investment strategies! Where the money goes is likely to be the biggest determinant of what research projects are available to you.

Working in a fashionable and therefore crowded but growing niche has advantages and disadvantages. The major advantage is that it likely is easier to get funding, and you might find it easier to get your work published in the higher impact journals in your discipline. However, those advantages are offset by there being a higher likelihood of someone else working on the same question as you and that leads to a danger that they may publish before you, making your study harder to get out. My advice is not to pick a project just because it is in fashion, base your choice on your interest first.

Choose a project which allows you to develop new and/or advanced skills

At different stages of your career, you should look for different things in terms of personal development. If your current thinking is that you will aim to pursue a research career in academia or industry, then your project choice should consider how the associated training will help you will develop into a better researcher. All projects should teach you how to do research at a fundamental level, things like; experimental design, effective record-keeping, data analysis and statistics, data presentation, publishing etc. However, the methods used to generate the data could be wildly different from one project to another. Similarly, your skillset will make a difference in terms of what jobs you can get and who will want to collaborate with you. When it comes to writing grants and moving upward into faculty-level positions, what you can *prove* you can do continues to be important. My advice is in the early stages of your career, look for projects where you will grow in some tangible way.

Big Tip

Your project is about more than the research question, it is also should provide personal development opportunities.

A big decision is whether you should become a specialist in a specific area or aim for broader development. An advantage of specialising is that you have definitive skills that can give you an edge in that specific area and allow you to do the most cutting-edge science. There are clear disadvantages though; it can become more difficult to change direction if you have become too narrow in your focus too soon.

My recommendation here depends on your career stage. If you are choosing a short (<6 months) project, then being focused in one area can be great. Ideally, choose a different area for each mini-project or lab rotation to get a well-rounded base. However, for a PhD studentship, I would want to ensure I included a few extra elements in the project to build in some safety. The additional skills don't need to be advanced techniques, they could be as simple as the preparation steps required to generate the samples, or the follow-up experiments to validate the findings. As a postdoctoral researcher, it is important to keep progressing. If I had specialised in my PhD, then I would look for something that would allow an additional layer to be added to that skill, or something to complement it.

If you do choose to specialise, make sure it is in an area that is expanding and looks like it will continue to grow in the future. You don't want to be the best at something obsolete.

Choose a project that balances risk and reward

For short projects high-risk, high reward is fine. However, if you are embarking on a PhD program, you want to be sure that whatever happens you will be able to get your PhD! Remember that negative data is just as important as positive data, and you can write up a doctoral thesis with either. The issue to try and predict is whether a high-risk project might not generate anything, could it simply not work? Before you panic about this possibility, lab heads do not generally design projects in all or nothing way! Usually, we build in some safety elements to make sure our students can get through.

As this an important point, I recommend asking about contingency plans during your interview.

For postdoc positions, the high-risk high-reward project could really accelerate your career. Nature, Cell, Science etc publication open doors. As a postdoc, you are likely to be in a position where you can develop smaller, lower-risk side projects in addition to your main work. By balancing your time and plugging away at these smaller projects whenever possible you can keep papers coming out regularly and offset any risk of the big project not going where you hoped. Again, a good thing to discuss before you start.

Supervisor and Group Choice

Another important thing to consider is who will be working with, especially your supervisor(s), and the people who will be training you in the lab. Again, your decision should involve identifying what *your* needs are both in terms of in lab training and in terms of support for other skills like writing and data processing. The more honest and realistic you can be about yourself, the better. Can you work on your own? Do your writing skills need work? How frequently would you want to meet with your supervisors? Knowing what you want will allow you to ask the right questions when you interact with potential supervisors. The next few sections are worth thinking about.

Danger: broad generalisations ahead! None of the points below are definitive. You will need to assess each team its own merits.

Established star, solid performer, or up and coming?

It can be very attractive to work for the superstar scientists. Of course, it is. They have got to the top because they are good at what they do. The top profs will have had more overall experience in supervising students, and usually their network of collaborators will be more extensive than a newly appointed faculty member. This second point means that you might have a better chance of meeting other prominent players in the field and that recommendation letters for jobs and

Big Tip
Get a feel for how you will be supervised by asking current students and staff how the research team is run.

fellowships carry more weight. But, be aware, that just because they have published in the top journals and supervised many students, does not necessarily mean that a superstar prof is actually a *good* supervisor. Almost inevitably, with more demands on their time, there will be less time available for you and this could mean you will be reliant on the other people in the group to help. Indeed, the prof may never actually set foot in the lab, and might not have done the most up to date techniques with their own hands. This could mean that they might have unrealistic expectations in terms of time-frames or challenges involved.

In contrast, a newer faculty member might still be research active or, at least, likely to have done the techniques that you are using more recently. Often a new faculty member is protected against teaching and administrative commitments when they first start and likely they will want to invest more time in you. It's much more important to a junior faculty member's career that a student succeeds.

Irrespective of the potential supervisors' career stage, while you are at your interview try to ask current students or team members how often they meet with the boss, how their training is managed, and then weigh those comments against what you think you will need.

My supervisor is amazing! 3 Nature and 1 Cell paper already this year!

Wow! What are they like?

Don't know, I've never met her!

Be aware that the superstar PI might not have much time for you!

Multiple supervisors (mentoring team)

Many institutions have moved to a model where more than one faculty member supervises each student. If this is the case, you might have an opportunity to offset any disadvantages of one supervisor with positives from the others. The second supervisor might provide a skill that the primary supervisor does not have, or it might be the case that the more junior faculty member provides the day-to-day guidance while the established professor gives you the high-level advice but may not see you as often. This dual-layer supervision can be quite a good compromise. It is not necessarily all good news though. If your team have different outlooks on where the project should be going, then you might find yourself having to manage different supervisor's expectations.

When you are at your interview, try to meet the whole supervisory team or, at very least, establish how the supervision will work. Will it be a clear primary supervisor with limited input from any secondaries, or will it be more evenly split with regular whole team meetings?

Group size, funding and focus

An additional consideration is how the team dynamic works. For small groups of under five people, you would expect most, if not all, to be pushing in broadly the same direction. In contrast, larger groups might have a hierarchal structure with a few senior post-docs or lab managers effectively running their own sub-teams. In either case, you need to know how *you* fit within that framework, who you will be working alongside day-to-day. Note group and individual work can both be great, one is not necessarily better than the other. You should consider your character within your decision. Do you work well with others? Would you prefer working on your own?

While you are at your interview or through the lab website and publication history, you will be able to find out what everyone else is working on, and then work out if you are going to be part of that team or if you will have a completely independent project. As a team, you might publish more frequently, but there could also be competition for who gets the most credit for the work.

Funding is vital, not only for determining the group size but also in what *type* of work you will be able to do. If you are considering a PhD studentship, it likely will include some bench fees in the funding. However, those may not be enough to do every experiment or technique that you will want. Therefore, find out what other money and equipment will be available to subsidise the major costs associated with your project.

You want to be able to do the right experiment in the best way, rather than being restricted to only the experiments you can afford (although, even the top labs still have to budget).

Publication frequency and quality

If you plan to go down an academic path in your career then primary data publications will be the most important aspect of your CV when it comes to future job and fellowship applications. Therefore, irrespective of your career stage, you should consider the outputs from a potential supervisor in terms the quality and frequency of original data papers.

On the face of it, it can appear that simply the more papers your supervisor publishes, the better. However, it is not quite as linear as that, you have to adjust for some additional aspects. First, it is probably best to ignore anything where the supervisor is not either listed as the first or last author (or co-first/last), as in every other location it is likely that they were only peripherally involved. Second, you should consider the size of the person's group; bigger groups should be publishing proportionately more frequently. Third, look for different first authors within the list of publications; you want to know that each lab member is productive and not just the star postdoc. Fourth, and probably most importantly, look for the ratio of more substantive primary data papers to more descriptive, lower-level papers. You want your supervisor to publish *good* work, not just lots of work (we'll discuss what makes a paper "good" in later sections of the book). Finally, you need to acknowledge that everything is field-specific. A clinical discipline might be able to publish lots of case reports very rapidly, whereas a cell biology or transgenic mouse model can often take years to come to fruition.

I appreciated that all of this isn't easy at early career stages, my general advice is to ask the opinions of some of your academic advisors. For example, when I was choosing my postdoc direction, I took a list of prospective future supervisors to my PhD mentor and we discussed the pros and cons of each. His insight was far deeper than I could glean and helped shape my decision.

Institution Choice

It's not just the person, the place matters too. Here are some things to consider.

Hype versus reality

You can do valuable research anywhere. However, there are reasons why the highest quality research comes out more frequently from the same Universities; success leads to more grant income, and more money leads to the potential to do bigger, more elaborate or fancier experiments. The fame and money also mean that those institutions attract a greater proportion of the talented scientists, meaning that the research environment might also be more conducive for better career development. This will lead to extra pressure to perform and you need to be careful that that doesn't translate into toxic levels of competition. Of course, you already knew all of that; however, it is important to be aware that all of these perceived benefits come with a heavy pinch of salt. Merely being in a higher tier school doesn't automatically mean that your potential project will be better nor that your supervisor will be good at training new researchers.

While I do think it is relevant to consider the quality of the overall research institute when making your choice, it shouldn't be the most important item on your list. In the grand scheme of things, there aren't actually many differences between the higher tier institutions, especially in terms of resources. What is most important is whether the appropriate financial and academic support and equipment are available for *your* specific project. When you go for an interview, don't be wowed by the shiny new building, instead, think more critically about project-relevant equipment and check the institution websites to see what core resources are available.

Location

One of the major benefits of science is that it is an international endeavour and we can work wherever we want. Moving between institutions and even between countries is not only common but also is seen as a positive thing on a CV. Indeed one of the real benefits of Science is that the work is truly multi-cultural, working and interacting with people from around the world is excellent for widening your world view and will make you better at your job. Of course, your visa situation. funding restrictions and your family situation may limit the choices available but, otherwise, you should aim to do your project wherever is best for your development and your specific research area.

It's OK to have life outside the lab!

If you are thinking about PhD programs, be aware that each country and even the institutions within a country will have different examination requirements and degree duration. This information will be available on the relevant webpages, so do some reading before you apply.

Training program

If you are choosing a PhD program, you should find out not just about the research project but also about the rest of the framework for learning. Find out about the other skills they teach. A good program will include training writing and career-related skills as well as other core teaching such as in statistics, experimental design.

Teaching and non-research commitments

Before you make your final decision, find out about the expectations of a PhD student within that institution. You may or not be required to be a demonstrator or teaching assistant for a class, and this may or may not be a good thing! Having a small amount of teaching experience on your CV may help some future job applications and supplement your income, but if you end up teaching a lot, you might not have time for research. As with the other things on these pages, it's all about knowing yourself and knowing your needs.

Stipend vs cost of living

Will you be able to afford to live? Hopefully, the answer to that is a definite yes, but will you have enough disposable income to be able to live *the way you want to*? The fantastic project with the outstanding supervisor in the top school may all come crashing down if you are miserable all the time or spend most of your days commuting because you can't afford to live near to your work. Some cities or countries are better or worse than others in terms of cost-of-living, so it is do some quick calculations and make an informed decision about what your lifestyle will be like before applying.

Application checklist:

- How interesting is the project (to you)?
- What skills will you develop?
- What other training opportunities will you have?
- What are the chances of participation in high-quality publications?
- How often can you expect to publish as part of this team?
- How do you feel about the supervisor and team? Will you fit in?
- Is the lab funded sufficiently for your project?
- Is the institution well equipped? Well respected?
- Where is it? Will you like living there?

You may have to make compromises, decide what is most important to you.

1.3 Applying for a Research Position

Now you've chosen the ideal position the next step is to actually get the job offer. Not a trivial endeavour, there is lots of competition out there. This section highlights some important things a prospective supervisor is looking for so that you can maximise your chances.

Informal enquiries

Big Tip
Use a formal email address for all correspondence (e.g. your University account)

If a job advert provides an email address for informal enquiries, then you should take that opportunity to make contact. Ideally, do this as far in advance of the application deadline as possible. Although ostensibly informal, these enquiries are a chance for you to make a good first impression and can help you stand out from the competition. Write a **short** email that states who you are, why you meet some of the core requirements of the position, and why you are interested in the position. I would also attach a short-CV or resume (1-2 pages). Remember that the person you are applying to receives 100s of emails each day, so keep this first contact message short and focused.

Example informal enquiry

RE: (position name) informal enquiry

Dear Professor ...,

My name is xxx, and I am a ... studying... at... (CV attached). I noticed your advert for a ... and it sounds like an exciting opportunity. Specifically, it interests me because...

Could you tell me... (a specific enquiry)? Would it be possible to have a brief chat about the post?

Many thanks in advance for your time.

Kind regards,

CVs, Resumes

Before you can get your dream position, you need to get a foot in the door. You need your application to be good enough to be selected to interview. For that to happen, your CV or resume, along with the accompanying cover letter need to make it crystal clear that you meet the requirements of the post and that you will bring desirable qualities that make you an attractive hire.

The purpose of the CV or resume is to deliver the *relevant* facts about yourself in as clear a way as possible. There are lots of templates available, but whichever one you choose, make sure it allows a reader to quickly and easily pull the important pieces of information they want without having to hunt for things. Organise the document logically with clear sub-headings;

- Name and contact details,
- Educational background with dates,
- Relevant work experience with dates,
- Relevant extras like awards,
- Grants, and publications if you have them.

If you are applying internationally, then relevant language proficiency and your visa status or right to work in that country should also be clear. It should never take a reader long to find these core things. The document should, of course, be free from formatting, spelling and grammatical errors. Check it thoroughly, then get other people to check it.

I'm sure whenever you have been told about CVs, the importance of tailoring it to the job has been reinforced. I agree. However, good news, the majority of your CV stays the same for every job. Job-specific adjustments can often be easily achieved through changing the length of sections to give the more relevant experience more space and reducing anything else. Another option that is really quick to achieve, is to change the location of the parts you want to draw attention to. Move sections around so that most valuable things are toward the top or middle of a page while less exciting things are at the bottom. Avoid spreading anything across page-breaks.

When it comes time to hitting send on your application, I recommend converting your CVs or resumes into a pdf. This will ensure that all your carefully crafted formatting decisions remain how you want them once opened by prospective employers.

Cover letters

Your cover letter serves a slightly different purpose than your CV. The letter is your chance to explicitly spell out why you want this position, to highlight the skills and characteristics you have to match the requirements, and what extras you can bring to the project that means you are better than the competition.

Big Tip

Assume that the person will not have looked at your CV before they read your cover letter.

Essentially the letter should tell them why they should interview you. The challenge is doing this in the most concise and focused way possible.

Structure

As with your CV, your cover letter needs to follow a logical path. Before you begin, make a list of what they are looking for (in the job description) and then identify what you can use from your background and experience to *demonstrate* that you possess those skills. If the list is long, identify points that can be delivered together, for example, combine skills in specific laboratory techniques or analysis platforms into single entities. Once you have that list, you will be ready to assemble the narrative.

The people reading the cover letters may have hundreds of applicants' letters to read. Therefore, you want to deliver a focused, complete, but concise message, that is easy to absorb and doesn't go on too long. Overall, t about one page of single-spaced text, possibly a page and a half is about right. Any longer than this, and you lose impact. I also suggest you opt for slightly shorter than normal paragraphs to help the letter feel direct and to the point.

In terms of structure, you have some options. Probably the easiest for the reader, and for you to write, is to establish broad platforms/concepts at the beginning of the letter and then focus on the specifics in later paragraphs. Example layout below:

Opening	~80 words*	State the position you are applying for and your headline qualifications / credentials
Why are you applying for this position	~150-200 words	What is it that attracted you to the position? What will it mean for your career progression? Be specific.
Highlight *essential* skills and experience	~150-250 words	Use the job advert to identify what skills or experience is required, then clearly and succinctly write about how you meet those primary criteria. Don't repeat anything you said in paragraph 1 or 2.
Describe *desirable* skills and experience	~100-200 words	As above. Draw attention to things that will help you in the role but which might not be obvious from your CV / resume.
Describe character traits and other relevant information	~100 words	Use evidence to support statements regarding hard working, productive, team player etc. If relevant, state your visa status, language proficiency or any other potentially relevant information.
Wrap	~30 words	Positive ending (no new information).

*Word counts are indicative only, but don't go on too long!!

After addresses and a salutation of "Dear Professor" or "Dear Selection Committee", the first paragraph should be quite short. State the job you are applying for and the critical details about yourself, such as your current position or what you are studying at present. Finish with an upbeat transition into paragraph two.

In paragraph two expand upon *why* you have applied for this position. Use some broad statements about your background and how that background connects to the post. Consider starting this paragraph by stating "I am applying for this position because…" this will keep you on track.

Paragraph three is the place for you to address the specific requirements of the job. The topic of this paragraph is something that implies, "I am appropriate for this position because." You can find the job requirements in the job description, or can infer them from the project description. You may require two or three paragraphs to deal with all the points depending on the length of the advert, but try to be succinct and focused. One way to introduce natural breaks between these paragraphs is to talk about specific skills or experience in one paragraph, and then character traits in the next. Remember that the details of your experience, are all in your CV. You don't need to repeat these in the letter, instead your aim is to show how your experience is relevant to this role.

Finally, end with a one or two sentence positive closing statement saying that you look forward to hearing from them.

Respect the Robots

One final point to be aware of; if you are applying to a big or prestigious program or large company, then the cover letters and CVs might be read by artificial intelligence as a first screen before a real person looks at them. The AI will search for keywords.

To make sure you have these keywords, aim to incorporate the same phrasing of key terms as they appear in the job specification to get through this first screen.

Evidence

Everything you write on your cover letter, or in your application, or that you say at interview should be supported by some sort of evidence. Backing up your statements is important for any job you apply for, but it is especially important when the people making the decision are scientists (we are all about proof). In your letter, a good mantra is "assert then justify"; make a clear statement of some specific information about yourself that makes you appealing to the selection committee then justify your ability to make that assertion by providing supporting information. Indeed, you should only mention things on your cover letter that you can support in some way.

Cover letter phrasing: evidence

Compare

"I am passionate about animal welfare."

With

"I am passionate about animal welfare and have volunteered on numerous occasions at the animal shelter."

The first sentence is something that anyone who applies to the job could say; it won't help you stand out from other applicants. Adding supporting evidence makes the statement believable.

Desire and Commitment

One of the traits that your assessment committee are looking for is someone who actually wants to do the job! Do you care? Will you turn up? Will you give up if things don't go well? As someone who sits on interview panels, I always believe that I can train people in specific research-related skills where needed, but it is much more difficult to teach someone to be hard-working. If you are able to demonstrate that you have the sort of determination and work ethic that means "things get done", then it will dramatically improve your chances.

Having evidence to indicate that you will care, that you will turn up and work hard, and will be productive can be tricky! Later in your career you can use your publication output as a sign of productivity; however, earlier on it's likely you won't be able to speak to that. Things like part-time jobs and involvement with sports teams and hobbies can help you're here. You can use these to support statements about commitment, dedication and collective responsibility, or to illustrate your time management skills. That bar/coffee shop/supermarket job may not sound relevant, but if you frame it appropriately, it can help your case.

Tailored

You might need to apply to many positions before you get invited to an interview, but churning out the same tired cover letter will not help you get past this first step. I'm sorry but you will need to spend a little bit of time

Big Tip

Tailor your application to the type of job, to the specific project, to the supervisor, to the dept. and to the institution.

making it specific to the position. If you are applying for a funded position, the person deciding who to appoint will have put time and effort in to get the funding, the last thing they want is someone who doesn't care enough about the job to take the little time needed to make their cover letter specific.

Always tell the selection committee not only why you are right for the position but also why the position is right for you. Where will it take you, what will you learn, why have you picked this specific project? There are lots of correct answers to these questions, so just be honest.

Interviews

Getting invited to interview is an achievement in itself. Now it gets really serious. At interview, you will compete with people who are likely to be just as qualified as you are. You do have reason to be confident though; the committee wouldn't have picked you if they didn't think you had potential.

The ways interviews are conducted varies from position to position. To alleviate some of your nerves, try to find out as much as you can beforehand about what to expect. For example, find out if they want a presentation, its length and topic.

Attending vs Zoom

Prior to the COVID-19 pandemic and social-distancing, I always recommended that attending in person rather than using video conferencing if at all possible. However, travel has become much more difficult and for the foreseeable future you are less likely to be disadvantaged. If there is an option to attend in person, and if travel is possible then I do think you should try to take it. Travel expenses should be paid, so if you are only talking about a few hours trip then deciding not to come would raise questions about how much you wanted the job. Choosing not to take an international or inter-state flight is much more understandable. One of the post-COVID advantages will be that committees will be much more understanding of a reticence to travel.

Be aware that it is harder to respond to people and pick up on non-visual cues if you are not in the room. If you do use video calling, make sure you do it somewhere with stable and robust internet and somewhere you won't be disturbed. If you are going to use things like screen-sharing to give a presentation, make sure you know how to use those tools beforehand.

If attending in person, you should do everything in your power to be on time. Check out campus maps, Google street view etc. so you know which building you are going to. Reducing uncertainties will help lessen any pre-interview nerves. Even the best plans can still go wrong, your transport methods could let you down, and the interviewers will understand this. If it looks like you *might* be delayed, then you should let someone know ASAP. Identify a person to contact who is not part of the interview panel (such as a administrative assistant) as the panel may be busy interviewing and unavailable.

Be nice to everyone

This should be obvious, but at every stage of the interview process, you should be courteous, polite and respectful with *everyone* you interact with. You should be like this always, of course, but it is especially important in interview situations as you don't know whose opinions the interview panel will solicit when making their decision. When I interview prospective candidates, I always ask my current staff or students to show the candidate around the building and to have a chat with them. This tour is for the candidate's benefit, but I also always ask my staff for their opinion too, they will be working alongside them after all. I also ask the receptionist for their comments. Irrespective of how well a person fits the science of the project, if they will not fit with my team, then I absolutely don't want to employ them.

Project Awareness

In your cover letter and CV, you (should) have emphasised project-relevant points, and unsurprisingly this continues and becomes even more important at the interview. You should do some background reading about the project and the supervisor before attending. It would also be useful to have some ideas on where the project might lead, where you would go next if everything "worked". At the very minimum, read some abstracts from papers published by the group on your journey to the interview. It shouldn't take much effort to understand the core principles so failing to do that reflects poorly on you. Where necessary, also have a read of some company websites to learn about techniques that you don't already know. If the project is related to a disease or tissue or organism, you definitely should know something about that topic area.

Remember, it will be impossible to convince someone that you care about a project if you don't know what that project is!

Presentations

At the interview, you might be asked to do a short presentation on your research background *and how they fit the job*. Carefully read what is requested in terms of length and content and stick to those rules. Especially important is not going over time, that is guaranteed to annoy the committee! A presentation is an opportunity to show what you will bring to the job without having to hope for an appropriate interview question.

A job-talk isn't the same as scientific meeting presentation, even when you are presenting your past work and using the same data slides. In a job talk, the faculty members want to know if you are right for *their* position. When I am interviewing someone, I want to know the *quality* of the data the person produces and the way that they think about that data. rather than the scientific story. To address your listeners' needs, you need to change the balance of the talk to focus on skills and data rather than (for example) providing a lengthy, detailed introduction or implications of your previous project.

Don't assume that the interview panel will be able to connect the dots between what you are saying and what they want to hear. Instead, you should explicitly connect your work with the requirements from the advert.

Example Interview Questions

Tell me about yourself, talk me through your CV.

Most interviews will start with some variation of these questions. This opening question is to help you settle in. As an interviewer, I want to you to do well in the interview. I don't want nerves to limit your performance. However, it is also not a trick question. The panel will have read tens or maybe even hundreds of cover letters and CV, this question helps to remind them who you are and is a chance for you to highlight some important things from your CV.

If you think there are anything that might cause concern within your CV, this is an opportunity to address them. For example, you could consider bringing up any reasons for gaps in employment history within this initial narrative. If you think something potentially negative will come up during the discussion anyway, it can be better to address it yourself and therefore control how it is framed.

Why did you apply for this *type* of position?

Sounds like a weird question, but the interviewers might want to know why you applied for a PhD rather than a technician or vice versa, or why a postdoc rather than a fellowship. We want to know the motivation for your career trajectory as it informs us about what you want to get out of the opportunity. Your answer should explain what this type of position will mean for your career and personal development. Many candidates for PhD positions answer this with some variation of the PhD being a "logical next step". That's sort of OK, but it does suggest that you may not have *really* thought about it. What is this a step to, and why is it logical? What will having a PhD mean for you? A PhD is a major undertaking and commitment both for you and for the supervisory team; if you don't have an answer to this question, you should ask yourself if you are really ready (and you know the interviewers will think that about you as well).

Why did you apply for this *specific* position?

This is a different question to the previous one. Here you should demonstrate that you know and care about the project. Be specific! Don't just say it is an interesting topic; say *why* it's interesting to you. Talk about the skills you will learn and how they will help your career. Talk about the strengths of the team you are joining and why you chose them. You can even talk about what it is that made you select the institution. This last point becomes more important if you are applying locally instead of choosing to move elsewhere. The job location matters, but generally, from an interviewer's perspective, it shouldn't be the only motivating factor.

What do you expect to gain from this position? Where will you be in 5/10 years?

This question is a staple of interviews for every job-type. You might cover the answer in one of the previous questions but your interviewers will almost certainly ask you to tell them about your goals. There isn't one correct answer to this question, but this question presents an opportunity for you, again, to illustrate that you have considered what this position will mean for your career development. Be honest; there is not a right and wrong answer. For example, if you see yourself doing an MBA, patent law, science writing or entering industry after your PhD, then you will want to indicate how this specific PhD is the best one for you. A short technician post to get a feel for whether you want to go on in academia or stay with a professional service career path are totally valid responses.

If you were designing this project, what would you do?

Design an experiment using (a technique that your CV said you have experience in) to answer a project-relevant question.

I ask variations of these two questions to see how the candidate thinks. I also use them as a mechanism to assess how honest they have been in their CV. If you claimed to have experience of using lots of different techniques but have only really shadowed somebody or followed a protocol during a practical class, then don't try and fake it, it will be obvious! If I think you are misrepresenting one thing on your CV, then it will be hard to belive anyting else. Importanatly, if you can't trust someone, you won't employ them.

If the job advert has mentioned a technique that you haven't done at all, then read a bit about that technique before you turn up so you can show that you want the job! Even if you haven't done *any* lab work before, this question is a chance to demonstrate that you have read the literature and absorbed how science is done. When preparing for your interview, pay attention to *how* the studies were performed so that you can suggest techniques for use in the future.

Any questions for us?

Every interview will end this way. There are lots of things that you *should* want to know the answer to at this stage. It's good to have something but don't ask questions just for the sake of it! Especially those that you could have found out by checking the job specification or lab webpages. One to three questions is about right in a short (30 min) interview, you can ask less critical questions via email later. Prioritize questions about the science over those about logistics, as the science questions give you another opportunity to demonstrate that you care about the project and have thought in depth about it.

Some example topics for questions you can ask:

- About the project: e.g. which journal will they target to publish, what techniques will you be able to learn, or more specific questions about the science involved.
- About the lab: how big is the team, how many other PhD students are already working there, how it is equipped, how is the team run? Also, ask if you can be shown around.
- About the training program and requirements of the program.
- About opportunities to teach or to attend conferences.
- About collaborations on this project or on how supervision will work.
- About life in the city/country, cost of living etc.

It can be a good idea to have a list of questions printed out so that you can refer to them. You might realise that all the questions you had have already been answered. Don't worry about this, it's likely a good thing. If you want to demonstrate that you came prepared, one option is to list off the questions from your list that they have already answered, "Well, you have already told me about... and ...".

After the interview

There are some cultural differences in what happens next. In the US, after an interview, it is generally expected that candidate should send a "thank you for interviewing me" email. This is less expected in the UK bur it never hurts to be polite. Even if it turns out that you don't want the job, it is worth sending a short email to thank the people for their time and the opportunity to meet with them. You never know who you will interact with in the future.

The decision about who to employ may take a few days, but, hopefully, the committee will have told you when to expect a response. You deserve to know either way, so if you haven't heard anything for about a week, it is appropriate to follow up by sending a short, polite query email.

You won't get every job you apply for

Don't be surprised if you aren't successful, especially at first. Interviewing well is a skill; it's something that takes practice. Even if you are well qualified *and* perform amazingly well at interview *and* get on well with the interview panel, you still might not get the job. Don't be downhearted by this, there are a myriad of reasons why they might have selected someone else, and it may be nothing to do with you. I can guarantee that every successful scientist has been unsuccessful sometime in their career. Just because you didn't get *this* job doesn't mean you won't be successful in future.

I know, from experience, it is frustrating when you don't get a position you think you would be good for, but make the most of the time you have invested by asking for feedback to improve your chances for the next job. Again, always be courteous and gracious in your interaction. The feedback you receive might make you better at interviewing but also might help you plan your next steps, for example, looking for work experience or doing an online course to fill a skills gap.

While self confidence can take you a long way, it is worth
having a plan for how you will answer some of the
standard interview questions.

1.4   Starting a New Project

Joining a new lab and starting a new project is exciting, but it is also can be intimidating. You have a lot of new people to get to know, new procedures to learn, a different administrative framework to get to grips with, inductions and training on new or different pieces of kit, as well as all the things outside of work that come with a move. On top of all that, there is the project itself to learn about, which means a lot of reading and experiment planning. It might be several weeks or even months before you lift your first pipette.

Here are my four recommendations for what to do first:

1.   Get to know the technical and administrative staff

You don't want your induction periods to drag on longer than is absolutely necessary. A lot of the rate-limiting things you need to do at the start will require input from people that aren't part of your lab team. Certain training sessions and inductions might happen on a schedule rather than on-demand, so when you identify what training is needed, don't procrastinate, get yourself booked in. Top tip; be friendly with the technical and administration teams, it will help you to get the boring but necessary stuff out of the way more efficiently both at the start and throughout your career.

2.   Learn what the rest of the team are doing (and have done)

Knowing how your project connects with the rest of the team will help you fit in. At the beginning you will need lots of advice; knowing who does what will help you identify the best person to ask your questions. You might find work a little slow at the beginning of a project, therefore shadowing the team when they are doing something new to you, or helping out when they are doing something that you are confident in, will get you out of your seat and feeling productive.

3.   Learn the background

A new project and new techniques mean you will have a lot to learn. Read, read and read some more! Don't forget, not everything has to be a journal article. Textbook can help lay a solid foundation, which will help you to absorb deep information. You can also mix it up by doing things like online training, watch YouTube videos or webinars. Finding information is a crucial skill, the next section looks at it in more detail.

4.   Read the grant

Most research is funded through a grant from a government or charitable organisation. To get awarded those monies will have required a grant proposal. Early in the project, you should become thoroughly familiar with this document. Anything you don't understand you should go and learn more about. This knowledge will help you to contribute more in all future conversations with your supervisors. Keep a copy of the grant and read it again after a few months when everything makes a bit more sense.

1.5 Information Gathering

The more you know about your field, the better you will be equipped to conceive and design experiments, and the easier you will find interpreting your data and writing about the findings. Before you begin working in the lab, your course or supervisor might require you to write a preliminary report or literature review. These are commonly asked for as they are valuable ways of ensuring that instead of passively reading the material, you actively engage with it. It's not just "busywork" to fill your days; knowing the literature will help with every other aspect of

It will help you with all your research to thoroughly understand the topic and to know what has gone before. Don't be surprised if your first tasks involve reading the published work and producing a literature review.

your project and is essential to becoming a good scientist. Even once you have the background on board, you should still be reading. I recommend scheduling time in your weekly plans to ensure you always are adding to your knowledge.

Note that at the end of a project, a literature review section serves a slightly different purpose; it tells *the reader* what they need to know to understand the rationale and significance of your work. Later in this book is a chapter on writing literature reviews for a thesis or for publication. At the beginning of a project, this exercise is more about you, it is about developing your personal knowledge.

Let's look at where the information comes from.

Learning the Field

What you need to know	Where to find the information
Fundamental understanding of core concepts	Websites, textbooks, popular science articles
Focused appreciation of the current thinking within the field	Review articles, scientific editorials and opinion pieces
Detailed knowledge about your study question	Primary data research articles
Methods and protocols in your model system	Primary data research articles, company websites, methods books and methods papers

Secondary Sources

Textbooks, websites, and popular science articles

Your first goal should be to get acquainted with the core concepts of the field. Hopefully, you will have had lectures or modules on the basics, if not you might want to see if you can take some now.

You may need to begin with some of the more easily accessible writing styles

Big Tip

It will be easier to read more detailed papers once you have a good grasp of the basics. Use textbooks and websites to get up to speed.

first so that you can read and understand the deeper literature more easily in the future. This initial learning you are doing is all about laying the groundwork. There are robust data showing that learning the same material using different methods leads to better long-term retention; therefore, use as many sources as possible, including video sites or other easily accessible mediums. However, be aware that when you write up your work, websites and popular science articles won't go deep enough for you to rely upon them as credible sources. Indeed, I **would not** expect to see a website or opinion piece cited in a literature review or final paper unless it was obtained from a source with verified quality. Don't cite Wikipedia!

Textbooks are usually better than popular science websites in terms of the quality of the material that they provide. Good books will have been through peer-review and multiple editions, and this means that the messages within are well-researched and are generally accepted by the research community. Books take a long time to get published so remember that they will not carry the most up-to-date material, for that you will need primary sources.

In your final write up, you can use books to support broad, general concepts but you shouldn't use a book as source for specific details. Indeed, if you are writing a primary data research paper, you usually don't reference books at all.*

Some core textbooks describing experimental techniques get cited frequently in methods sections. Check your PI/lab manager's bookshelf for the "classic" texts such as Maniatis "Molecular Cloning", Harlow and Lane "Using Antibodies".

Ah... a night with the lab-classics.
There's no finer night-time reading

Big Tip
Review articles can help you learn the field and also can help you to identify the important papers from the ref list.

Review articles

Once you have a grasp of the fundamentals, your next step is to go a little deeper and a little more up to date. Most journals publish review articles. These are often commissioned articles that are then written by eminent scientists in the field where they summarise current thinking on a specific topic or present their new ideas. It is likely your supervisor will have written some recently, make sure those papers are in your reading list

Review articles don't generally contain new findings but instead combine the conclusions from multiple recent publications with the previous thinking that led up to them to provide a holistic view of an aspect of the field. In many ways, a review article is similar to a literature review, except the article form is usually shorter and more focused and tells a story.

The better review articles go through the peer-review process where other established scientists in the field check that the authors have been fair and comprehensive. However, peer-review only goes so far. Do not forget that most* review articles are *opinion* pieces, and there may be other interpretations of the data being discussed. Therefore, you should read multiple reviews on your topic written by different people. You want to get a broad feel of different opinions in the literature.

In your written assignments and manuscripts, review articles can be used to set up key points. Limit your use of review articles to occasions when you are making broad rather than specific comments. If you are using a review, include *"as reviewed in"* in your in-text citation to indicate to the reader that you are not citing the original information source (more on how to cite later in the book).

In addition to improving your core knowledge, reviews are also great for identifying important papers within your field. Every subject area has certain papers that have made a big difference and are cited in almost every review article. These frequently cited papers are the ones that you should read next.

*A *"systematic review"* is a slightly different entity – it involves setting up a clear approach (a system) for deciding which papers to include or exclude. The writing style feels a bit different.*

Primary Data Publications

Books, reviews and websites, are "secondary" sources. These publications do not report new data, but rather present ideas or summaries based on previous observations. Once you have a good grasp of the core literature, you move on to using the original reports of the research, the "primary"* sources. These will contain the most up-to-date findings. It is particularly important that you read primary sources in rapidly moving fields, including life sciences, where much of the information that is relevant to your study has not have been written about in reviews yet.

Big Tip

Just because a paper has been through peer-review doesn't mean that the conclusions are now "facts".

Not all primary data publications on your topic will be useful to you; many will be of poor quality or limited in scope. So, how do you decide what to read, and, ultimately, what to cite? There are different strategies you could use to cut down your reading list into manageable chunks and to make sure that you read the papers that will be most useful to you. Here are some tips:

Read your supervisors' papers

This is an obvious but important place to start. Anything that your supervisors have published that is relevant to your project is worth reading. In fairness, anything that is even peripherally related to your work is also probably worth looking into. In addition to the core knowledge of these papers, you should pay attention to the writing and data presentation styles that your supervisors use, likely this will be what they expect from you.

Ask for a reading list

You might find that your lab has collated together a list of the core papers in your area into a reading list. Ask around, if this exists, then awesome! If it doesn't, think about creating your own so that the next new lab member knows where to begin. In the grant funding your work, your supervisor will also have cited some essential resources. Go and read all of those.

Read lots of abstracts

The abstracts of papers give you the central story elements of the paper in short form. You will not get enough details to discuss the paper in any depth, but from the abstract, you can get a good idea of what they found and what approaches they used to come to those conclusions. If they've done a good job in writing their abstract, you will also know the motivation and implications of their findings. Put some keywords into an appropriate search engine (more on this shortly) and then take some quick notes on abstracts identifying which ones to go back to for a more comprehensive read later. Performing a quick pre-screen will help reduce the amount of time you spend reading non-relevant work.

Through reading lots of abstracts you should be able to identify the main points you will need to cover in your intro, discussion, and lit review. However, I want to emphasize that only reading the abstracts **is not enough** for you to judge if the findings are supported or not. You have to read and critically evaluate the whole paper.

This is also a good time to point out that once you come to writing up your work for publication, you will (of course) need to write an abstract. As you will appreciate from your paper hunt, spending extra time making sure that the abstract for your paper is as good as it possibly can be is really important. It makes a big difference in terms of how many people go on to read your paper in depth and how much of an impact your work will have. Later in this book, we will go into abstract writing in more detail, but as you will be reading lots of abstracts at this point, try to identify which of the writing and style features of a good abstract makes you more likely to go on and read the full paper.

Consider the type of study: observational, mechanistic, or interventional

If you are working in a medical discipline, you likely will have been taught about the hierarchy of research evidence. Usually drawn as a pyramid, this places secondary research into an approximate order of how confident you can be about the results. This hierarchy places meta-analyses and systematic reviews at the top; both of these use large bodies of data to increase their sample size and thereby make more robust

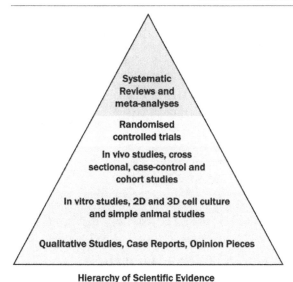

Hierarchy of Scientific Evidence

inferences. At the wide base of the pyramid are expert opinions and descriptive type studies where the data may be anecdotal, case reports, or otherwise observational with small population sizes.

The pyramid is designed for clinical research but you can think about papers from lab-based research in a similar way. At the bottom of the pyramid are papers that observe or describe a phenomenon but do not go any further. The next step up includes papers that try to understand the mechanism behind phenomena, while the top stages may put that mechanistic information into relevant biological contexts. Reading all of these different papers *will* add to your body of knowledge but they each play a different role in your development.

Where you should spend the most time, i.e. what type of papers you should read, depends, in part, on what you intend to do in your studies. If you are planning an observational study, then you should be aware of every observational study to have gone before to be sure that you don't repeat anything unnecessarily. In this case, you might get less benefit from going deep into mechanism and intervention-type papers (at least early in your time in research). As your research moves up the hierarchy, the emphasis shifts. If you are trying to identify an interventional approach, then you likely should spend most of your time reading mechanistic and interventional procedures-type studies to know those details in depth.

Of course, you should never stop reading. Staying on top of the literature is an essential part of being a good scientist and you should try to read everything. However, considering the type of study might help you to prioritise your reading lists.

Consider the model system and how it relates to what you will be doing

Just like considering the study type, you should consider the relevance of the model. It's usually quite easy to identify the model system(s) used in a specific published paper; it will be indicated in the title and/or abstract.

It might be beneficial to focus your earliest efforts in systems that are close to the one you will use. What I mean here is that if you are planning to work on protein interactions in solution, you might not get much specific, project-relevant value from studies carried out in a whole organism or vice versa. Once again, to be clear, knowing as much about your project as possible at every scale will enormously help you to interpret your data in the long-run and you will need this info for your discussion, but, in the short term, in the first weeks in the lab, focusing your reading closer to your destination will provide the most instant value to you for planning experiments.

Use the citation count of an article (relative to its publication date) as an indication of its importance

Citation counts are visible in websites like WebofScience and Scopus (see "finding information" below). Once you have a list of papers that look interesting, considering their citation count divided by the time since they were published.

A paper that has been cited by many other papers is likely to be more important or influential than one which has not been cited at all. Indeed, if other

> **Big Tip**
>
> Scholarly search engines will usually allow you to see citation count during your searches.
> You can search for a keyword then sort by citation count to quickly find the "big" papers.

people cite a seminal paper regularly, it will be strange if you don't. Read that paper and see what the fuss is about. It may not be great, but you need to know about it!

Use the journal of publication as an indicator of quality

Each discipline has a series of journals that are more difficult to get into than others. Being published in a top journal does not *necessarily* mean that any individual paper is particularly valuable to the field, but as a method of prioritising your reading list, it is worth considering where the work was published. There are many ways in which journals are ranked (see box below), what is "good" is very much dependent on what you are working on and the relative size of the field. The rankings also change over time. These systems aren't great. For example, we all know that the most commonly used ranking - impact factor - isn't a good indicator of a paper's quality. However, we currently do still look at the ranking lists when choosing where to submit our manuscripts! You'll also find that your colleagues have opinions about how well/poorly a journal is edited.

> ### Journal ranking systems
>
> There are thousands of journals publishing scientific work. Different ranking systems have been set up to help identify which journals make the biggest impression on their field. None of these ranking systems are perfect and all should be used with caution!
>
> **Impact factor (IPF) and Citescore**
>
> These two ranking systems use the average number of citations per paper in a journal in either a 2-year (IPF) or 3-year (Citescore) window. All publication types published by a journal count toward the score, including review articles. This is important, as reviews from the big players in a field tend to attract more citations than primary data articles. Therefore, a journal with high review to primary data ratios will have artificially elevated IPF.
>
> Using IPF is usually reasonable to identify the top tier journals; a journal with IPF above eight is likely only to accept good or controversial papers. Either way, at least the work will be worth reading! Similarly, anything published in a journal with an IPF score

below one is something to be wary of. There will be individual good papers in these low impact journals, but you will have to look for them and judge them on their own merits. In journals with IPF in between 1 and about 8 (field dependent) you will find lots of good stuff but also lots of rubbish! You really have to make your own judgement on a case by case basis.

Mega journals like PLOS and Scientific Reports publish tens of thousands of papers a year. IPF and Citescore metrics are meaningless in this context as some papers may receive thousands of citations while many will receive none. The mean calculated from those thousands of papers doesn't tell you anything about an individual paper you are reading. You really have to make your own decision.

Eigenfactor scores (www.eigenfactor.org) **and SCImago Journal Rank (SJR)** (www.scimagojr.com)

These two metrics rate the total importance of a journal based not only on the number of citations but also on *where* those incoming citations come from. This is a way of balancing out the relative size of the field. Journals publishing in popular areas (e.g. cancer) might have higher IPF as there are more papers being published and more cross citations, wheras journals in smaller fields (e.g. eye research) but which make more meaningful impact within their might have a lower IPF score, but in the latter case they would have a higher Eigenfactor score. If you work in a smaller niche, using the Eigenfactor score might give you better feel for a journal's standing.

Google h index (h5 index)

The h index is usually used to measure an individual's outputs rather than to compare journals. The h index is a single score per person that measures the number of papers (h) that have received (h) number of citations. An h index of 10 means that that person has ten papers with at least ten citations. The h5 index is the journal equivalent system; except, in this case, it is limited to the last five years. An h5 of 100 means 100 papers have had at least 100 citations, although many other papers may have received far fewer citations than that. Mega journals will have much higher h5 scores than their impact factor might suggest as they will have many papers with high citation counts.

Article-level metrics (Altmetrics)

You won't find Altmetrics as a list for each journal, but whenever you click on any individual paper, you often will see the altimetric colour wheel on the webpage. This little infographic tells you the number of citations, reads, downloads and other info like tweets as news reports for that article. Although not at the journal level yet, Altmetrics are growing in popularity as a way of rating the impact of an individual

> paper. There is a clear benefit, they are accurate for the individual paper and they capture real-world interest as well as academic citation counts.

Preprints; brand new but not peer-reviewed

Almost all the papers you read will have been through a peer-review process, i.e. assessed by other eminent researchers in the field. The peer-review process isn't perfect, but it attempts to make sure that only those studies that have been designed well and whose conclusions are supported get published. If you ever get asked to review a paper, take it seriously; you are shaping the world's knowledge. However, in addition to the fully reviewed work, there is a growing movement for pre-publication manuscripts to be made available. The flagship server for these preprints are currently bioRxiv (www.biorxiv.org), and medRxiv (www.medrxiv.org) for biology and medicine respectively.

Preprint manuscripts have the newest data in them, and it may be worth looking for any specific papers in your topic area in case other people have answered the questions you plan to ask but haven't published them yet. You must remember that these manuscripts are not thoroughly reviewed, and therefore, the evidence may not be as robust as it first appears so be extra critical of the information.

Additional sources for methods, protocols and advice

Almost all the experiments you will design will be based on modifications of those that have gone before. This is good; you don't need to invent everything from scratch. Textbooks and journal articles are great sources of protocol information, and there are also specific journals that publish methods-style papers. Company websites selling kits or reagents for specific experimental types have improved massively and many now provide useful pages on the design and approach to performing experiments. Of course, they will also try and sell you items, but likely their sales pitch will help you identify aspects of your experiment that you need to consider.

When you are working in the lab, you will undoubtedly come up against problems at some point. Whenever you are struggling, remember that someone else will have very likely have had a similar issue. The science community is pretty good at helping each other (we are trying to make the world a better place), and there are lots of online forums where you can post questions and can get ideas for solutions from the community. These forums can help you spot an obvious mistake and can help you get back on track quickly.

Finding Information

Search engines Google, Bing, and Yahoo etc.

I'm starting here as you already know this; type in your keywords into a search bar and go for it. Search engines' algorithms rank the pages depending on a variety of metrics related to the number of clicks, inward and outward references in addition to the match of your keywords to the content of the page. This means that they will return an artificially ranked list that elevates the more popular pages. This might mean that the material that is actually the most relevant to you is hundreds or even thousands of pages down. To get around this, use Boolean search terms (AND, OR, NOT etc.) or surround phrases with "..." to narrow your search.

Google Scholar, ResearchGate, F1000 etc.

To get around the problem of search engines returning non-reputable sources, one option is using "Google scholar" rather than Google. Check it out, it is free, and the using the same search terms within this more curated collection will often give you more relevant information. ResearchGate and Faculty of 1000 (F1000) are other amalgamation sites, almost like social media for scientists, they each have slightly different approaches but both allow you to search within their databases using standard search terms.

Libraries

Remember them? Seriously, libraries are a great source of material including not only textbooks but also all other forms of scholarly writing including journal and electronic articles. In the past, journal holdings were centralised in the university library, but now almost all journals have online archives. However, many journals make you pay to access their publications so you might still need your university's library (physical or online) to get what you need. The library can also arrange interlibrary loans of the specific material that you can't get your hands on. Your library almost certainly has a website that will have instructions on how to make use of their journal subscriptions by logging in remotely. Learn how to use these tools, especially if you ever plan to access papers outside of your institution (i.e. at home). Good libraries have trained librarians who are some of the best people around at hunting down information. They'll be more than happy to help you get that obscure but potentially seminal paper from the 1950s although beware their penchant for keeping places quiet!

Finding the seminal papers was beginning to feel like a bit of a quest

Subject-specific databases; Scopus, PubMed, WebofScience

While the previous methods all work, the best way to focus your search onto the scientific literature is to go to one of subject-specific databases. For the sciences, www.webofscience and www.pubmed.com are likely your best bet. The built in tools allow you to narrow your search to reviews or primary literature and sort by year, author etc. You also can see related text and articles that have cited the paper you are reading. You can (and probably should) set up alerts to let you know when new papers are published including your keywords. These databases will be where you will spend the most time when hunting for papers. Scopus www.scopus.com has similar functionalities for searching papers but it also has some other added features such as article and journal-level metrics, which can be useful for gauging quality and deciding where to publish your work

Big Tip

Scientists want their work to be read! It's OK to ask for a copy of their paper.

Direct from the author

Modern science is rapidly moving toward an open access publishing model where everything is available to the reader for free. However, some publications, especially older ones, will not be available to you or your library due to paywalls. You likely will have an option of paying individually per article, but those costs are really expensive. Instead, you can ask for a copy from one of the authors. Look for the corresponding author's email address around the author list and addresses part of the article page. Asking the author for their paper used to be very common, and you will find it still works as a strategy. As always, be polite and patient in your requests.

Note Taking

As you are reading papers, you are going to want to take some notes. You can do this in different ways and hopefully you have an idea from your studies of what works for you. In addition to paper notes, there quite a few free programs and apps you can use. I like to link all my notes to their sources, so apps, where you can do that or at least copy and paste links work well.

You want your notes to help your writing. If your records take longer to read than would take to read the paper itself, then they aren't worth having. On the next page are my recommendation for where to start:

Find a method of note taking that works for you!

Take Notes

- Identifiers: Title, Authors, Journal (copy and paste).
- Study type: e.g. Descriptive/observational, mechanistic, interventional, Meta.
- Approach: model system and methods overview.
- Key findings: bullet points of conclusions.
- Ideas for how this paper relates to your work.
- Rating: how good is this paper? how valuable to your work?

The last two points are the ones that will help you the most. It will force you to think about the work you are reading and this will help you to retain the information. This type of thinking takes practice, but it is an important skill that you will need for experimental design, writing grants and for interpreting data.

Pay attention to writing styles

The more you read, the more you will become aware of the differences between good quality papers and those that aren't so good, both in terms of the Science and the writing. One of your goals in becoming a top-level scientist should be to produce high-quality work. Therefore, you aren't only reading papers to learn the information, you should also pay attention to the writing and presentation aspects of the paper. Specifically, consider the way that the "story" is delivered and whether that is effective, i.e. does it make it easier for you to absorb the information? Look for themes and patterns in how introductions and discussions are assembled. When looking at data figures, try to identify what "works" in terms of layout, labelling, assembly and what feels professional.

Later in this book we will go into detail about each of these points. For now, keep aside some papers that are *written* and *presented* well (irrespective of the impact of the science) to use as project-specific examples.

Chapter 2: Experimental Design

2.1 Project design
What you should consider when planning the whole project.

2.2 Experimental Design Preliminaries
What you need to think about before you begin, including formulating testable hypotheses.

2.3 Correlative or Manipulative
How to choose between the two main types of experiment and understanding their limitations.

2.4 Measurements
Defining the data you will generate and how it will be captured.

2.5 Controls
Identifying which extra samples will be required to improve the quality of your results and to rule out alternative interpretations.

2.6 Independence and Randomization
Defining the experimental unit and removing bias.

2.7 Sample Size Calculations
How to calculate how many samples you will require, plus a gentle introduction to some statistical terms.

2.8 Final Checks
The last things to think about before you start collecting data.

Resource
At the back of the book, there are "Experimental design checklists" these will take you step by step through the experimental design process.

2.1 Project design

The Scientific Method – Hypothesis testing

Before we get into the details of experiment planning, a quick reminder of how science works.

Projects start by identifying a problem or problems that would be *valuable* to solve. This problem could be a gap in the knowledge about how a particular process works, it could be trying to determine the explanation for a series of intriguing observations, or it could be more tangible issue like trying to treat a disease or solve an environmental problem.

Once you have defined the main research problem or question your next job is to come up with potential solution or answers. To do this, you draw on everything that has been done before, any preliminary studies that are available, and discussions with your colleagues (this is why a project always starts with reading). Often you will come up with a list with multiple potential answers to your question. The next step is to decide which is the most likely. This becomes the *central hypothesis* of your study. The project will be designed will test this hypothesis. If there isn't enough information to form a clear hypothesis, then you might need to start by doing a "hypothesis-independent" or "hypothesis-generating" experiment, more on these shortly.

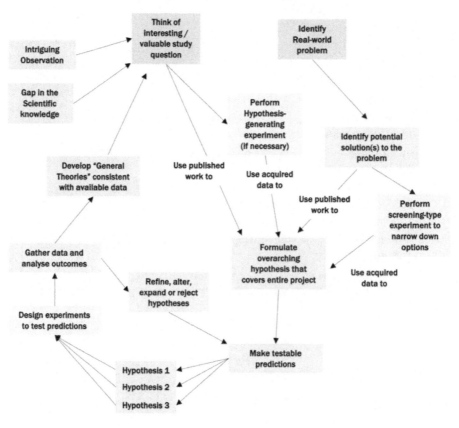

Usually a central hypothesis is too wide to be effectively tested using a single experiment. Therefore, you next break the question down into a series of smaller, more specific questions each with their own hypothesis that you *are* able to test. You design experiments around these *experimental hypotheses* and the results will combine to support or refute your central hypothesis. With discovery science-type projects, the project is very likely to shift as you begin finding answers. As you gather data and learn more about the system based on the findings, you will

> **Big Tip**
>
> Science is about asking questions.
> Don't ever be
> afraid to ask for help!

update or modify your central hypothesis, and then design new and better experiments to test this updated model. The findings may generate new questions or allow you to progress to the next stage of a problem-solving type project.

Note: often the long-term project plan will already exist when you start your project. It may have come from your supervisor as part of the grant application; however, as you grow as a scientist, you should expect to be more and more involved in project planning.

Step 1: Identify the big problem

Start with the big questions. What is it you want to know? What question would be the most *beneficial* to answer in terms of opening new avenues of research, adding to the body of literature or solving a problem.

Write this question down. It will help you to focus if you have a clear overall project-level objective in mind.

Hypothesis-generating experiments, fishing trips and big data

Sometimes it is not possible to come up with a central hypothesis. This could be because there isn't enough information to be able to predict the answer, or it could be that the potential answers to the question are so wide-ranging that it would be impossible to effectively address them all. At this point, it might be beneficial to do a *hypothesis-generating* experiment. These types of experiment are essentially observational studies where you record many things about your subjects and look for associations or correlations between an observation and an outcome. They could also be an initial screen of large compound library trying to find *any* hits that could be the next drug. Early experiments like these often provide descriptive (qualitative) readouts rather than quantitative measurements. However, most of the experiment-level design aspects that follow in the rest of this chapter apply equally to hypothesis-testing and hypothesis-generating experiments.

One area that has grown massively in the life-sciences is using Big Data and "omics" approaches such as transcriptomics, metabolomics, and proteomics to either generate hypotheses or, less commonly, to test them. These approaches allow researchers to identify *everything* that is different between populations. The experimental question would be like; "in which ways are population one different from population two?" or "what are the global changes induced by treatment Y".

These types of experiment can be really valuable as they allow you to look at things in an unbiased way, without preconceived notions. They require tight, focused design if you are going to get the most out of your work. The biggest challenge they present is that by looking at everything at once, you inevitably will get lots of *false positives*. During the design phase, decide how you will validate what is real as opposed to what is a spurious finding. Usually, this means that after you do an -omics-type of study, you must follow it up with a series of more focused hypothesis-testing experiments. Where you go after experiment 1 should be part of your project plan. There may not be value in generating data if you can't validate it.

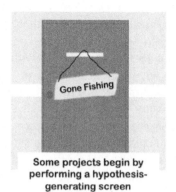

Some projects begin by performing a hypothesis-generating screen

Fundamentally, the thought processes remain broadly consistent for big data experiments compared with more focused experiments, so read on. However, you do have some extra things to consider. The technologies are advancing extremely rapidly and new recommendations for how best to design these types of experiment are published regularly. Therefore, rather than cover them here, I strongly recommend looking up the most recent requirements and seek advice for experimental specifics.

Triangulation

No experiment is perfect. All experiments have limitations, sources of error and variability, and other potential confounding factors or scope for alternative interpretation. Good design minimises these problems and increases your confidence in your inferences. However, the best way to improve the reliability of your findings is to ask the same question in multiple different ways. We call this *triangulation*.

As we will discuss shortly, a smaller and more focused experiment gives results that are easier to interpret and require simpler statistical methods. However, smaller experiments will not capture every aspect of the project-level central

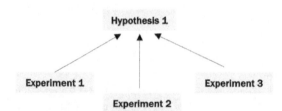

It's unlikely you will be able to design a "perfect" single experiment to test your hypothesis.
More likely you will need multiple lines of individually imperfect evidence before you can make robust conclusions

hypothesis. A project rapidly becomes a jigsaw puzzle with each experiment providing a piece. By taking the composite results of each of the different parts and analysing these together, you will be able to offset any individual issue and thereby gain a better understanding of the whole. For bench science, this usually involves using a different technique or approaches to answer a separate piece of the overarching question. For example, you might choose to use western blotting and ELISA as complementary approaches to determine protein abundance. If the different methods each yield similar results, then you can be more confident in the overall interpretation. Triangulation is also used to rule out alternative interpretations.

When you come to write up your data, the discussion part of your research paper you will combine the multiple individual pieces of experimental evidence into a coherent whole. Indeed, this one of the key purposes of a discussion section.

Validation studies

A quick but important point. Most research involves finding out "new" things, pushing at the boundaries of what is known. However, this isn't the only valuable work that you can do. "Validation studies" are really important. It much like the triangulation point above, the more data generated that supports a viewpoint, the more robust the inferences can be. Validation studies are crucial for the long-term scientific understanding and must be reported irrespective of the outcome for the benefit of Science.

Validation isn't just about the whole project. At the beginning of your project, you may need to repeat some previously published work in your system to make sure that everything is consistent with the literature. Ask anyone who has worked in the lab for a while, and you will find they have a story about a reagent or model that they found doesn't work as reported. Test everything yourself!

Plan the whole project

Time management is a crucial skill which you will develop as you progress through your science career. It's not just a matter of making the best use of your time on a day to day basis to maximise productivity but is also about prioritising your focus within the framework of the whole project.

Big Tip
Spending the time to plan your project fully will be better in the long-term. It will get you to your goal more efficiently overall.

I recommend planning for *all* the separate sub-questions of your project near the beginning (at least as an overview). Your goal in this planning is not about deep details yet, but rather it is to identify the steps which will take the longest and which experiments will generate results that will influence future directions versus those that will provide an endpoint in themselves. Making a rough calculation of the total time it will take to complete each part of your project will help to ensure you don't run out of time. If you are grant funded, then often these timescales will be laid out in a Gantt chart (see below).

For each of your project's sub-questions, think about what preparation you will need in terms of making new reagents, generating or validating model systems, applying for ethical approval, breeding of experimental animals, recruiting patients, or growing up your crops, bacteria, cells etc. Basically, anything that could take a long time (see box below on *rate-limiting steps*). Additionally, try to estimate the difficulty and likelihood of encountering problems during optimisation.* Use local experience to help you with this. Finally, consider how long the actual data collection part of the experiment;

The preparation steps for an experiment can end up taking a lot longer than the experiment itself. Making sure that everything works properly is critical to getting robust data

this stage will be dependent on the number of times you need to repeat the experiment or how many participants you need to recruit (we will come back samples sizes later). It may not be possible to get accurate numbers for all these items until you have a pilot study, but you can make some estimates based on published work. Always build in some wiggle room into your timeframe, things inevitably take longer than you anticipate. Even a perfectly planned experiment has the potential to be delayed by illness, reagent availability or delivery delays, or other unexpected events. The longer the experiment, the more variability there will be.

Most lab-based studies actually involve far more time in set-up and optimising than in collecting usable data.

Plan the paper

At the beginning of a project, it can be beneficial to formulate a plan for how the best-case scenario version of your paper(s) will look. If you are doing a short project, then your work is likely to be a small part of a more extensive publication rather than a separate entity. However, your supervisor will have a plan for how all the work will come together, and it is beneficial for you to have an appreciation of that plan. Of course, the plan is likely to evolve in response to the answers your experiments deliver, it is a living document; however, having a good idea of where you are trying to get to can help you to focus your energies upon the right area.

Examples: Rate-limiting steps

In school, you probably came across the phrase "rate-limiting step" in Biology or Chemistry. In terms of project design, these are the steps that will determine how quickly your overall project progresses. You might prefer the idea of doing some fun experiment in the lab, but often it will be better for you to focus on something else. Identify the things that will slow you down and make sure that those are always a priority that doesn't get forgotten. Here are some examples:

Animal studies

If you are generating constructs to make a transgenic animal, then all the cloning steps might require in vitro validation before you start work with the model organism. Depending on the model system, you may need ethical approval before you can begin. Will in vitro validation or the ethics will take the longest? If it is the ethics, you should do that first, even if it is hard, new or boring.

Once you start the animal work, you will have breeding steps and the ageing of the animals before you can start. If you are bringing new animal lines into your facility, then there may be quarantine and acclimatisation requirements to wait as well. Also, you might have personal training and licensing requirements to consider. Each of these steps *could* take a long time, differs from project to project, and need to be factored into your planning.

Human studies

Ethical approval will be required before you can enroll participants. However, to apply for ethical approval, you will need to have a complete study design. So, completing that design and writing the ethics application should be your #1 priority every day until the application is in. Once you have obtained approval to begin, the rate at which the participants are enrolled into the study is likely to be the biggest determinant of the time to completion. Focusing your energies on those steps rather than on anything else will help you proceed more rapidly.

In vitro studies

Things generally can happen faster in in vitro studies, but likely you will be working on more than one experiment at a time. Again, you should think about which experiments will take a long time. For example, the 3D model might take months compared with 2D analyses, so should be starting with the more complex but also more time consuming part? Priortise parts of the study requiring the most optimisation, and those where you need to prepare new constructs or reagents.

Project planning: Gantt charts

A standard project planning approach is to create a Gantt chart (named after the inventor). This is a sort of bar chart with the different project stages on the Y-axis and the predicted time to completion on the X-axis. A Gantt chart also indicates where there is dependency between the different stages, i.e. you identify rate-limiting steps. The more complex, more multi-faceted the project is, the more important these plans are. Many grant applications require a Gantt chart as part of the application, so you might find one already exists for your project, although you will probably want to add more detail.

Originally people would draw up Gantt charts with pencil and paper (still a valid approach), but you can use Excel or PowerPoint just as easily. In addition, there are lots of dedicated project management programs and Apps available. The newer management systems can be especially useful if working within a large team. Search online, and you will find a wide range of options or templates to get you started.

Creating a Gantt chart

- Identify each experiment that will make up your entire project; they become your first subheadings.
- Within each experiment, identify all the steps that are required.
- Order those steps based on dependency; which steps must be completed before the next step can begin.
- Assign time to each step based on all the information you have available. Be realistic here. Things usually take longer than you would like.
- Factor in time for vacations and holidays.
- Include time to analyse the data and to write up the results. These are essential steps to a project.
- For a group project, assign individuals to different tasks.

Identify the rate-limiting steps and prioritise them.
Be realistic; things take longer than you expect

2.2 Experiment Design Preliminaries

The overall project plan gives you a framework; however, the really in-depth planning occurs at the level of the individual experiment.

> **Big Tip**
>
> It is better to answer one question fully than get incomplete answers to three questions.

Experiments can vary dramatically in scale from massive multi-centred, randomised controlled clinical trials which take several years to set up and run and involving thousands of patients spread over a whole country or continent, down to a protein-to-protein binding assay performed in a test tube which may only take a couple of hours from start to finish. Although the size, scale and complexity of the study make a difference to the planning, the core concepts are always the same. The next few pages will lead you step-by-step through the critical questions and decision points involved in designing your experiments. My goal is to instil good practices which will shape the way you think about all experiments, your own and in published work.

Experiment planning is important. To do it well takes time but it really is worth it. Poor experimental design leads to uninterpretable data and wastes time, money and, especially for animal or human research. is ethically unforgivable. Well-designed experiments also require much simpler statistical methods to extract useful information. Spending a little more time in the beginning, makes your life easier in the longer term.

There are three types of experiment:

- Ones which yield interesting data, no matter the outcome (good).
- Ones which generate interesting data, but only if it turns out a certain way (not ideal).
- Ones which yield ambiguous, uninterpretable results regardless of the outcome (bad).

Your goal should be to do everything in your power to ensure that you are doing the first type of experiment!

You can do this

You should go into your experimental design with a can-do attitude. Experimental design is ~50% about your understanding of the science, ~40% common sense, and perhaps ~10% an appreciation of statistical approaches (this should not scare you, we're going to make it simple). Of course, you can and should get help if you are uncertain about any aspect of your design. Also, most experiments will not be totally "new", they will be based upon of something someone else has done. This means that there will be examples in the literature* and that will make your planning much easier.

Remember that the published work may not be perfect, if you can improve on the design, you should

Experimental Design Preliminaries

A common problem I have encountered as a supervisor, is researchers conducting experiments without having a very clear and definitive question. The experiment feels like a "look and see" rather than the much more precise "I want to know…". This is equal parts surprising, frustrating and concerning from a supervisor perspective! If you do not have a clear experimental question, then your experiment will not be designed appropriately. It's as simple as that. Therefore, every time you start planning a new experiment, the first thing you should is write down the question you want to answer, being as specific as possible. Keep the question at the front of your mind throughout the design process.

There are examples for every aspect of being a scientist in the literature. Step 1 in any experimental plan is seeing what has been done before (then improving on the design).

Has something similar been done before?

Most experiments are, at the very least, inspired from work that has gone before.

Before you begin, do a thorough literature search looking for similar types of experiment. Ideally, you should look for examples published in the better-quality journals of your field. When reading these papers, take careful note of what the authors have included in their set-up with special attention to which controls they included (more on controls shortly). Always remember that these experiments are not identical to yours, but the things that they have considered in their design will be things that you, should also think about. Keep these papers close to hand so you can refer to them again as you progress through the design steps.

Form a hypothesis that answers the experimental question

Unless you are embarking on a descriptive (hypothesis-generating) experiment, now is the time to write a statement that definitively predicts what you expect to see. What is you predicted *answer* to the experimental question? Your experiment will test this answer. Whereas the project-level hypothesis statements are too big to be tested directly, the experiment-level predictions need to be tight. Each prediction should be the most robust possible test of your hypothesis. You want to walk away from the experiment confident that you have tested the hypothesis as far as you possibly can.

> **Experimental hypotheses must be testable**
> **Experimental hypotheses must be specific**

Include in your hypothesis statement all the specific details of what you will assess and what you will compare. Think carefully about the real-world implications of your prediction. If you are looking for a difference between populations, then both the *direction* and the *size* of the difference should be in your hypothesis. If you are looking for no-difference between groups (e.g. absence of side-effects of a drug), then your statement should explicitly state that there will be no difference. The numbers of samples and

Big Tip

You may need to return to your hypothesis and make changes later in the experiment design process

statistical tests you will perform reflects experimental hypothesis in terms of what you need to test and the effect size you are looking for.

Your hypothesis statement should also be specific in terms of the experimental approach. The procedures being used will determine the limits of how far you will be able to interpret the data.

Examples - Hypotheses

An effective way to spot a hypothesis that *isn't* specific or testable is to think about what the result would look like if your experiment refutes the hypothesis. What could you interpret from those data?

For example:

- *Treatment of epithelial cells in culture with drug X increases the expression of clock genes.*
- *Patients with disease Y have lower motor function than age and gender-matched controls.*
- *In a transgenic mouse model of Z, mRNA the expression level of X is increased two-fold compared with littermate controls.*

Each of these sound-like legitimate hypotheses. However, think about what would happen the experimental results *did not* support the prediction. Could you rule out alternative interpretations?

Are you planning to test *every* clock gene, *every* type of epithelial cell, *every* muscle function, or *every* mouse tissue? The answer to those questions is likely to be no. Therefore, these statements need to be more specific.

Something like:

- *Treatment of corneal keratinocytes in culture with 100 nM drug X increases the expression of the clock genes Bmal1 and Per2 by more than two-fold compared with vehicle control.*
- *Patients with disease Y have 30 % lower grip strength than age and gender-matched controls.*
- *In a transgenic mouse model of Z, mRNA expression levels of X in lung tissue is increased two-fold compared with littermate controls (you may even need the age and gender of the animal).*

Plan the Figure

When you start a journey, it always helps to know where you are going!

Early in the planning stage, I recommend making a quick sketch of how you envisage the figure in the manuscript, report or thesis will look.

Having a solid idea of you intend to plot on a graph, or the images you will need etc., can make it easier to answer some of the other design questions that will follow and will ensure you collect the correct data.

Again, refer to the literature as a starting point, remembering that your experimental questions may require a different presentation approach.

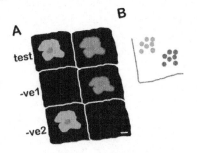

Having a plan for what the figure will look like will help you identify the data that you will need

2.3 Correlative or Manipulative?

Now that you have a hypothesis statement for your experiment, you are ready to start making decisions about study design. The first branch point is deciding whether to do a *correlative* or a *manipulative* study. A correlative study (as the name suggests) attempts to determine if there are correlations or associations* between different aspects of the study population. In contrast, in a manipulative experiment, the experimenter deliberately changes one or more experimental variable, and the impact of that change is measured on one or more outcome variables.

A project is likely to have both. Usually correlative studies are done first as observational studies used to study things in their native environment. Manipulative studies will then follow to test the hypotheses that have been formed in a controlled, closed system. Both experiment types have limitations. The best papers will often contain both correlative and manipulative experiments to provide a complete story.

Choosing between Correlative or Manipulative experiments

Correlative studies

Example hypothesis: "Smokers have lower lung capacity than non-smokers."

Use correlative studies when:

- To establish if associations exist in their real-life context.
- To establish hypotheses for later manipulative studies.
- If manipulation is impossible (practical or ethical reasons).

Manipulative studies

Example hypothesis: "Lung capacity increases after stopping smoking."

Use manipulative studies when:

- Mechanistic insight is required.
- You need to control for third variables/confounding variables.
- Reverse causation is a concern.

Technically: "association" refers to any type of relationship between two variables, whereas "correlation" refers only to linear relationships.

Considerations for Manipulative studies

Is manipulation possible?

Usually (not always) manipulative studies will allow you to use stronger statistical tests, and will yield results that are more definitive in their interpretation (X causes Y, rather than X is associated with Y). However, there are lots of times when it is infeasible or impossible to manipulate the system in the way that you want to. For example, when you are studying a long-term outcome, or when it is ethically inappropriate to intervene. A manipulation experiment telling people to start smoking in their 20s so you can determine if they get cancer sooner than those who don't is both ethically wrong, and might take 30+ years.

For many studies, the tools you will need to manipulate the experimental variable will not yet be available. For example, there may not be an appropriate knockout or transgenic animal line available or perhaps there isn't a chemical inhibitor of your target pathway. Many projects start by establishing a new experimental system or making the tools needed. Determining what to do is a balance between the value of answering the question against the time and cost of generating the experimental tools.

Can you produce biologically realistic manipulations?

In correlative studies, you usually look at things in their natural environment. This is an important benefit, you are studying the true biologically-relevant variation. Whenever you introduce a experimental manipulation you might be making your subject to do something that they wouldn't ever normally do or creating a scenario where the subject experiences an event but in a non-relevant setting. You must be careful in assessing what you will and will not be able to interpret from a manipulation experiment i.e. whether it will give you the information you are looking for.

This question of biological relevance of manipulations comes up for experiments at the cell or protein level experiments as often as it does for a whole organism. For example, if you drive overexpression of a protein with a viral promoter, you may end up with higher expression than would ever occur in real life. Similarly looking at protein-protein interactions in a solid-state binding assay, while interesting, won't necessarily tell you that the proteins interact in the same way in a cell.

How do you decide if your manipulation is realistic and meaningful? The answer, of course, depends on the question you want to answer. In the protein binding experiment, you would be able to determine if your protein was *capable* of interacting with other protein, but you couldn't say that your protein *does* interact. This is where the power of triangulation comes to the fore. You could combine a binding assay showing the capability of interaction with, for example, a cell staining-based approach to identify co-distribution and therefore the feasibility to interact. Together, these two experiments would allow you to make a stronger inference than either in isolation. Recognising what you can and cannot interpret from an individual experiment is vital when planning the whole project.

Importantly, if you recognising that the only manipulation options you have are not truly relevant may mean that modification of your experimental question and hypotheses is required. Remember, your hypothesis *must* be testable and specific.

Will your manipulation lead to any non-intended effects?

If they are non-intended effects, so how will you know? Answering this question requires you to use your knowledge of biology and common sense. Almost everything, be it a drug, miRNA, energy drink or whatever, will not be perfect in terms of specificity. There will be unintended additional consequences of the manipulation. Therefore, you need to carefully consider what off-target or non-specific effects your treatment will induce then try to account for these issues by using control treatments, (more on this shortly). If you are unable to design-out these potential problems, then you might not be able to answer your question the way you want, and, again, may need to backtrack and refine your hypothesis.

Considerations for Correlative studies

Correlation does not imply causation

You've probably heard this statement thousands of times. The biggest weakness of correlative studies is that you cannot be sure if the two observations are causally linked, or if there is some other reason for observing similar trends. You should continue to think about this point throughout the rest of your experimental design and identify ways to strengthen your study by choosing appropriate groupings and controlling for confounding variables. Ultimately, the best proof of a cause and effect relationship will come from manipulating the system; either preventing or inducing a thing to happen to ascertain their direct contribution to the outcome. This is why the best projects often include both correlation and manipulation.

Beware spurious correlations

Just because two things *appear* to be related to one another doesn't mean that they are. First, the association may be random chance (due to sampling variability). This is where statistical tests will help you. After your experiment, you will test for the likelihood of obtaining a false positive; i.e. of a seeing a correlation where none exists. The more things you measure at once, the more likely you will observe false positives. Therefore, when you are dealing with big data sets, where you have literally thousands of outcome measures, then it is **guaranteed*** that at least some of those associations are not real.

*Note, that one of the things you will calculate here is a false discovery rate (FDR).

Sometimes you can approach this issue of measuring too many things though using a tiered approach. You would gather data from a first group of samples, classed as a *discovery cohort,* where you perform a really wide analysis with many outcome variables. The results from these samples would provide a list of many correlations and trends. You would then use that list with a second set of samples, a *validation cohort,* where you ask a more focused question with fewer variables and stronger statistics.

The second type of spurious correlation is where the association is real, but the relationship isn't between the two variables. If X correlates with Y, and X also correlates with Z, then Y and Z will also correlate with each other statistically, but there may not be any direct correlation. For example, people in senior faculty positions at Universities are disproportionately male, and the average age of the senior faculty is in the 50s or 60s. Male pattern baldness occurs in about 50 % of males over 50. Therefore, if you compared baldness incidence in the general population against prevalence in senior faculty, you would identify a statistically significant correlation between being bald and being a professor; however, shaving off your hair would not improve your chances of being promoted.

In this example, there are ways to improve the design of the experiment to avoid making the spurious connection; using age and gender-matched comparison groups rather than comparison with the general population. However, sometimes you cannot use experimental design to avoid spurious correlations. This will either mean that you

Beware spurious correlations.
To establish a true cause and effect relationship usually requires manipulative study

need to be extremely cautious with your interpretation or that you will need a second experiment to rule out incorrect associations.

Sometimes you have to accept that if the compromises you will need to make are too great, then the initial correlative experiment may not be worth doing. Don't be afraid to step back and refine the study question.

Correlations do not indicate directionality

Even when the correlations are real *and* even when there are clear cause and effect relationships, there is another potential problem…will you be able to know the direction of the effect. Is it A that influences B, or is it B that influences A? This termed *reverse causation* and is something that you must consider in the design of your experiment. For example, body mass index is inversely correlated with the number of minutes of exercise per week, but does this mean people with higher BMI exercise less or that people who exercise less have higher BMIs? If the experiment only measured BMI and activity, it would not be possible to tell the direction of the relationship with certainty and you would have to acknowledge the potential alternative interpretation. Ideally you can adjust the design of the experiment to rule out potential reverse causation, but if you cannot then you may need to modify your experimental question to ensure that it will yield valuable data.

Choosing your model system

Model system choice is a trade-off:

- Biological accuracy vs complexity of interpretation.
- Biological accuracy vs ease of manipulation (*tractability*).
- Biological accuracy vs time, ethics and cost of the experiment.

Choosing the most appropriate model system for your question is an important part of the design process. Model systems can range from isolated proteins in a tube, 2D, 3D cells in a dish, ex vivo or post-mortem tissue, through to whole organisms and even discrete populations. This decision makes a big difference to every other aspect of your design and ultimately decides what you will and will not be able to interpret from the findings. The decision will sometimes be obvious, but don't fall into the trap of choosing the use same system you always used without first carefully considering if it truly is the best system for the specific question you want to answer.

As always, the key to deciding is having a clear question. You should always choose a system that recapitulates all the aspects of the biological system that you need, rather than automatically selecting the most complex system available. For some hypotheses, you will only be able to test the questions in the context of the full organism or tissue; however, for others, those complex environments might not allow you to unambiguously test your prediction. The decision is a compromise between biological accuracy versus the ability to manipulate what you are interested in, and also between the advantages of true physiological system against how easy it will be to interpret the results.

At the forefront of your model-system choice should be research ethics. Is the added benefit you get from using animal or human subjects worth the cost to the subjects? Often you need preliminary experiments in a simpler system to justify the decision to use the more complex model. Be aware that adding complexity might not only make your experiment longer, more expensive and more difficult, but may also not allow you to get a definitive answer to your question. You might harm animals or human subjects without even getting your answer. That is ethically unforgiveable.

The final decision about experimental work on animals or humans isn't a decision that you can make on your own; you will need to apply for ethical approval before your work can begin. Ethics applications vary from country to country, but irrespective of the local rules, the process involves making a case for why the experiments can only be carried out in the system you have selected as well as providing detailed information about the experiment has been designed to minimise pain and suffering and to yield the strongest, most robust, data.

Well, I've certainly got the model part sorted

It's inappropriate to use yourself as the model system

2.4 Measurements

Now that you have decided on your research question, formulated a hypothesis and chosen a model system, it is time to hone in to the specifics of what you will measure. Remember the hypothesis defines the data you should collect, not the other way around.

Here's what we need to think about:

- What experimental variables are required? Timings of treatments, number of doses, etc.
- Are your measurements direct or indirect, and what will that mean for data interpretation?
- Will you generate descriptive or numerical data, are there ways to convert observations into numbers to allow statistical tests to be performed?
- Will you need a pilot study?

What experimental variables are required?

What is it that you want to test? Most experiments test how the *dependent* variable(s) relate to or change in response to an *independent* variable(s). You need to consider all sides of the experiment; what treatment(s) will you perform, how will you break up your study cohort, and what outputs will you measure and how.

> ### Terminology: independent vs dependent
> **Independent variable:**
> A variable that stands alone and isn't affected by other variables. Age, genotype, or drug treatment groups. In practical terms, these are the things that you have chosen to compare. If you are picturing a graph, the independent variable will be along the X-axis.
> **Dependent variable:**
> The outputs you will measure. The things that will change in response to your independent variables. In the final graph, the dependent variable will be on Y axis.

Note that how you phrase the experimental question and hypothesis determines which variable is which:

Examples:

- If the hypothesis was *"the expression levels of gene X determine the crop yield of a plant"*, the independent variables are the different levels of gene X expression while the dependent variable is the crop yield.
- However, in an experiment to test *"the size of a plant influences gene X expression"*, the independent variable is the different sizes of the plant, the dependent variable is the gene expression.

Number of variables, cost and sample size

Next, we need to consider how many groups and how many variables we want to test. For example, how many treatments, how many timepoints per treatment, how many different doses.

It is always tempting to test everything at once, but if you don't *need* the extra groups to test your hypothesis, then you should seriously consider what value the added comparisons will bring. I want to reiterate an important point; answering one question well is better than answering three questions incompletely.

If you think you will want to compare many different variables, consider ranking their importance in terms of value in answering your question. If one variable is more important than any other then this becomes your *primary outcome variable*. Your statistical analyses will be focused on this outcome. However, if you can't pick one then you should be aware of the effect having multiple variables will have on your experiment.

As you add more and more comparison groups, you reduce the statistical power that each repeat of your experiment contributes. Or, to put it another way, for every variable you add, you will need to increase the sample size.

> **Big Tip**
>
> More variables mean more samples required, and more cost. Don't add extra variables if they don't represent real added value
>
>

At this point it is worth having a rough idea of what numbers it would be possible for you to process both in terms of cost per sample, sample availability, and the time needed to process each sample. Each of these may limit the number of different variables you are capable of testing and it might mean you need to drop some variables You also may find that you need to drop some variables once you have calculated their impact on sample size (more on this later).

If, after carefully considering this, you decide that you need to compare lots of things, then read up on *multiple comparison testing* before you start.

Direct or indirect measurements?

Next, I want you to look at your dependent variables and ask if they will directly test your hypothesis as you have written it or not. Examples of what this means are on the next page. If you plan to directly measure the variable, great! No worries, move along. If not, you will next need to decide if your indirect measurement system will be good enough to generate an unambiguous answer.

Indirect measurements are (pretty obviously) not as clean as direct measurements in terms of interpretation, but they can still have benefits that make them preferred choice in certain situations. Usually, these relate to cost, time or ethics. The concern with indirect measurements is that whatever you have measured could change in response to your independent variable for a reason other than what you want to study. If you can identify potential for problems before you do an experiment, then you can adjust the design of the experiment to account for alternative interpretations. Usually this is done through including extra control treatments.

If you do decide to use an indirect measurement, be sure to acknowledge the limitation in your write up and limit your conclusions to statements that can be supported by the data.

Example: Indirect vs Direct Measurements

Animal research

Glaucoma is an ocular disorder where the optic nerve degrades, leading to loss of vision. To measure the degradation of the optic nerve in an experimental animal requires killing the animal and processing the eyes for histology or electron microscopy. These are *direct* ways of measuring how advanced the disease is in different treatment groups. However, glaucoma is often associated with an increase in intraocular pressure (IOP). IOP can be measured in live animals in a non-invasive way, that doesn't involve killing the animal. Therefore, you might choose to use IOP measurements and obtain time course data information. IOP here is an *indirect* measurement of glaucoma; it doesn't directly tell you about optic nerve damage. As the experimenter it is your job to decide whether this measurement is good enough for you to test your hypothesis (taking into account what is accepted by your field of study). The best option is likely a compromise; make IOP measurements while the animals are alive and then use post-mortem processing to validate your indirect measurement. This is another example of how triangulation can improve the robustness of an observation.

In vitro assays

For cells in a dish or samples in a tube, most measurements can be direct. This is one of the major advantages of choosing a simpler model system. However, still think carefully about this to make sure it really is the case. A commonly used system for analysing cell viability is the MTT assay. In this assay, cells metabolise a substrate leading to a colour change. It is a quick, cheap and easy to do assay and in most situations, the amount of colour change is proportional to viable cell number. The assay is a *direct* measurement of metabolic activity but is often used as an *indirect* measurement of viable cell count. When used for viability, you need to acknowledge that the metabolic activity could change for reasons other than cell viability, and your interpretation would have to be more conservative. You should consider triangulating your MTT assay data with some other viability measure to allow you to be confident in your interpretation. The MTT part of the experimental series could be the assay chosen for high throughput experiments, testing more experimental variables. Whereas the slower/more expensive triangulating experiment might be smaller scale, done purely to validate the findings.

Descriptive or numerical data?

Part of the decision around what you are measuring is the *type* of data you are going to generate. If it is numerical, great, there will be an appropriate statistical test which you can look up, and the analysis should be straightforward. The data could be *binary* (yes/no, diseased/non-diseased), *nominal* (discrete bins, groups 1, 2 or 3), *ordinal* (ordered discrete options; low, medium, high), or *continuous* where the numbers could be anything (more of these shortly). In lab-based science, most of your experiments will be quantitative like this. Sketch the graph that you expect to obtain, and you are ready to move on. We'll use this graph again when calculating the sample size required.

If you have no intention of quantifying population data, then what you are doing is *qualitative* or *descriptive* research. This can absolutely fine; qualitative research generates depth or nuanced information about the topic, and it is often the qualitative work that produces the observation that establishes new hypotheses. However, if you intend to draw inferences about what your results mean, you will need an indication of whether your data are representative of the whole population. Or simply, how confident you can be that the findings are "real". To be able to make these assertions, you will need numerical data.

If you are planning an experiment that will generate imaging data, then plan how you will turn the images into numerical data

In science research, most data are numerical as standard. The common example when this is not the case is where the output is an image, for example microscopic analysis of cell or tissue staining. If the data aren't inherently numerical, you should explore whether there are any fair or accepted ways in which you could convert the images, gels, blots etc. into numbers. Put yourself in the position of the harshest critic of your work and ask yourself: "how can I prove that my "representative" example is genuinely representative of the population?" Whatever proof you come up with is most likely the data that your experiment could be designed to analyse quantitatively.

> **Big Tip**
> Plan your analysis approaches Before you begin. e.g. how will you convert descriptive data into numbers?

Ideally, the approach you use will be objective; it will generate numbers without human input but, if it is accurate and fair, then a subjective scoring system can be effective. To make a system as fair and as robust as it possibly can be, you must design in controls to remove human bias, e.g. by using blinding so that the people scoring the images don't know which experimental group they belong to (more on this shortly).

Do I need a pilot study?

Pilot studies are smaller-scale versions of your primary studies. You use them to ensure that every aspect of your experiment will work consistently, to nail-down the specifics of the procedures, to identify where variability will come from and how to minimize that variability, and to be sure that your measurement systems are as accurate as they can be. The answer to the question of whether you need a pilot is almost always yes!

Pilot studies are small scale versions of your full experiment. They help you establish protocols and are needed for sample size calculations

In lab-based research, more time is spent optimising protocols than generating final data. This is time well spent, a fully optimised experiment that consistently performs well will give you tighter data and need fewer repeats to yield robust data. Pilot experiments will save you time, effort and money in the long run. There is an adage: if you put rubbish in you get rubbish out. I also want to emphasise that even if you are using a reagent from published work or a kit from a reputable company, you should still do a pilot experiment to confirm that it will work as expected in your experimental system.

Pilot studies aren't only for lab-based experiments; things like surveys and participant instruction packs may need to go through multiple iterations and tests before they work well enough to use them in the field. You don't want to get halfway through your data collection before realising that you are missing important information. Any change you make will require amendments to your ethics so you certainly should test everything about your experiment before you go forward into the real subjects.

In clinical trials, you cannot change outcomes after you have started, therefore it is even more essential that you are 100% sure that our outcomes will work before you begin. When you are hypothesis-testing, pilot studies serve an additional role. Your pilot studies provide the numbers you need to determine the required sample sizes:

Pilot studies should identify:

- How big a difference you can expect between your study populations? (the effect size)
- How much variability will you expect *between* individual assays?
- How much variability you can expect *within* each assay, and where that variability comes from?

These values help you to decide how many *experimental repeats** and how many *technical repeats** are needed. They might save you time, money and effort as you could find you need fewer repeats than you might otherwise have thought.

Pilot experiments are not always possible due to the rarity of samples, research ethics or the cost to process samples. In these cases, you should try and optimise as much as possible to maximise your chance for success, and after that you have to rely on publications where others have performed similar types of experiment the give you estimates of numbers you need for sample size determination.

Experimental repeats refer to how many times you do the whole experiment, technical repeats are the number of times you measure each sample within the experiment. The difference is "independence" and will be discussed in section 2.6.

2.5   Controls

Controls are extra samples or treatment groups that you build into your experimental design to help data interpretation or increase confidence in your findings, or to check if your experiment is working and aid troubleshooting. We've touched on some of the uses for controls already, here we will go a bit deeper into these important parts of every experiment. You can classify controls into four groups:

Controls

- *Experimental* controls; to help with troubleshooting and confirm that the experiment has worked.
- *Biological* controls; to confirm that a positive result is really positive and that a negative result is really negative.
- Controls for alternative interpretation(s), for third or confounding variables, and for reverse causation.
- *Calibration* controls; standard curves to convert measurements into real-world units. Or internal normalisation controls to account for biological and experimental variation (including loading controls).

Your experiment can only ever be as good as your controls.
In many experiments, you will include more controls than test samples.

Experimental controls

Most lab-based experiments ~~can~~ will go wrong in one way or another. When they do, you will want to know why. The role of *experimental* controls is to help identify which steps are the source of the problem. Knowing this helps you to troubleshoot and adapt your protocols. You should always include your experimental controls, not only in your pilot and optimisation studies but also in your real experiments so that you have something internal to refer to in case you obtain some confusing outcomes.

Note that the experimental control samples get processed as usual but are only there to confirm that your experiment worked. They don't generate new data. Indeed, you already know the result from these samples. Therefore the results don't need to enter your statistical analyses.

Positive control: a sample that will definitely give data in this experiment

This is a sample that you are already sure will give a signal or a measurement in an experiment. That signal doesn't have to be interesting, it just has to work. If this sample doesn't give you the signal you expect, you immediately know something has gone wrong. This extra sample is something like a protein sample, DNA or RNA that was isolated in previous experiment and has been processed already, and you know is of an appropriate quality. A positive experimental control could also have been produced by a different mechanism, e.g. using plasmid DNA as an experimental control for a PCR or purified protein for a western blot or ELISA. The more you know about this sample, the better.

Positive and negative experimental controls help you troubleshoot problems. The first question you will always be asked about an imperfect experiment is how your controls performed.

Your positive biological control (next section) can serve as an experimental control, but this can be a bit risky in multi-step experiments. Indeed, you should aim to include a control for every step of your experiment, then, if there is a problem, you can quickly tell where along the pipeline it has come from.

Negative control: a sample that will give you a negative result at each stage of your experiment.

This sample follows the same but inversed line of thinking as a positive experimental control. The purpose of this negative control is to determine whether the signal you have obtained is real or if it has come from some artefact or mistake in the protocol. In practical terms, the negative controls are the easiest to include. Often all you need to do is use a blank well in your plate assay or replace one key reagent with water. Because they are easy, there is no excuse to leave them out! Again, controls for every experimental stage are important.

Example: Experimental Controls

Let's say you are planning a quantitative PCR (qPCR) experiment comparing the abundance of two mRNAs between untreated cells and cells treated with a new drug. This experiment has multiple stages; cell growth, drug treatment, RNA isolation, reverse transcription and qPCR for both the target transcripts. If you were to run the whole experiment without controls and got zero signal from the qPCR, you would not know where the problem was, it could have been at any stage. Therefore, we need to add experimental controls.

To identify what you will need, you look at each stage and think what can go wrong. In this experiment, you can look at your cells on the dish and should be able to can tell if they are sick before you begin; probably no extra controls needed here as you would cancel the whole experiment if you thought there was a problem at this point. The drug treatment bit will be controlled through biological controls, as discussed below. However, the RNA isolation could go wrong for purely experimental reasons, for example, you might have an

---

imperfect protocol for the cell type, or the reagents could be off or simply a genuine mistake could be made. Therefore, you should include a positive control for this step; an RNA that you already know is OK. Next, the RT-PCR step could fail for some reason (e.g. PCR machine not working), so in the following step, you would include a cDNA that you already know is OK. Once you reach the end of the experiment if your experimental samples give no signal, but the control cDNA and control RNA samples were both OK, you can infer that the problem has come early on. Whereas, if the control cDNA works fine but the control RNA did not, then you will know it's the RT-PCR step that has failed. Knowing where the problem lies allows you to fix it.

Biological Controls

Your experiment is only as good as its controls (I know I have said this already, but it's important). Whereas your experimental controls are included primarily for your benefit, the biological controls are required to confirm that your *results* mean what you think they do. These controls are critical for the interpretation of your data. When you write up your experiments, your readers will want to know what controls were run and will need to see the results from the biological controls to be convinced that your conclusions are fair and accurate. They go in the figures and in the stats.

You can never prove a negative result without a positive control

Most experimental hypotheses are that either there is a specific difference between the study populations, or that there is a correlation between the independent and dependent variables. The experimental outcome is either that your hypothesis is supported by the data **or** refuted by the data. This means you need to think about both situations.

The big question to ask yourself is: if you **did not** observe a difference or correlation, could you be sure that it wasn't a *false negative* result? Part of countering false negatives comes from running repeat experiments and performing statistical tests. However, you also must consider whether your experimental system would even allow you to observe a difference if it were there to be seen, i.e. is the experiment sensitive enough. One way to be sure it was possible to see a difference is to include a sample that you already know will show the effect that you are looking for. Unsurprisingly, we call these samples *positive biological controls*.

If the above sounds similar to the experimental control, that's not surprising. The subtle difference here is that the biological control is all about the *effect* you are trying to observe. It should be a real sample rather than a dummy sample. The example below should help:

Example: Positive Biological Controls

You want to test a new drug that you hypothesize will increase cell proliferation. When you run the experiment a lack of an effect of the drug treatment could be because the drug doesn't affect proliferation (true negative); however, it could also be because your cells were sick, were too confluent, or were already dividing at the maximum possible rate and therefore no further increase was possible. The drug *may* work but it wasn't possible to detect the effect (false negative). Therefore you need a control to make sure that none of these situations confound the interpretation of your results. You would include a positive biological control, a compound which has already been shown to increase cell proliferation. This wouldn't necessarily have to be a drug; it could be a growth factor or some other agent, just something that is already known to affect your model system (+test it in a pilot study). Now when you run your experiment, if your positive control shows an increase in proliferation and your new drug doesn't, you can be more confident that it is a true negative.

You can't prove a positive result without a negative control

The equivalent concept holds true for negative controls; you need to be able to *prove* that the result measured is driven by your test treatment and not by some other confounding factor. There are different levels of biological control, and you should aim to be as stringent as possible. Consider everything about your experimental set up that could influence your results but would not be due to the thing that you are testing, then design controls to account for those potential issues. Often, you will find that you need to include multiple negative controls.

Big Tip

Plan the figure!
The data from your control treatments should be included in your manuscript

The simplest negative control might be an untreated group, or wild-type animals (these will often be a comparison group anyway). However, as your treatment or interventions become more complex, then so too must your controls. Let's explore this with a couple of examples:

Examples: Negative Biological Controls
Example 1: Control treatment

Let's say you are going to inject your patients (or cells/animals) with a test compound that needs to be diluted in a solvent such as DMSO. You want to know what effect the test compound has, but when you measure the outcome variable you won't know if it is the compound or the solvent that has had the effect. The answer is to include a negative control group where you have a population treated with the solvent alone. This solvent-treated group will become your comparison group. Note that this sort of control is needed in lots of experiments, you always need to consider the effects the delivery agents could have. If you are using transfection reagents, viruses, or even just any buffer along the way, make sure you control for it.

> **Example 2: Transgenic animal studies**
> When you are working with experimental organisms where you have manipulated the genome in some way, then you need to consider every permutation of how they could come together. A common example would be inducible transgenic mouse models where expression of the enzyme Cre recombinase is driven from one promoter, and that enzyme then removes a "flox stop" cassette from a transgene driven from another promoter. You likely will need more than one negative control group: i) a control group with the transgene but without the Cre recombinase to control for the transgene being leaky or having landed in a functionally important region of the genome, ii) a control group with the Cre recombinase but without the transgene, controlling for any Cre-driven effect that is independent of the transgene, iii) If your enzyme was drug inducible, you might need even more groups; each of the comparison groups with and without the drug.

Placebos

You've probably heard of the *placebo* effect; people get better even though they haven't received the real treatment, simply because they believe that they have been treated. It's surprisingly powerful in a lot of situations.

A placebo is simply a negative control. Just as in lab-based experiments, the placebo should be as "real" as possible, as close to the complete treatment as you can, even if that means making an incision or administering something via injection. There are ethical implications for this; you must justify the potential harm caused.

Ideally, the trial should be *double-blind*. Both the person receiving the placebo **and** the person administering the drug or placebo should not know what they are receiving. This helps to control for unconscious bias in behaviour (more on this shortly). Be aware that merely knowing that they are a participant in a trial will unconsciously influence a person's behaviour; they might try harder in whatever task you set them. Designing ways to reduce these effects is not always simple, but doing so will make your data much stronger.

Pick your control treatments with care!

Special case: using antibodies

Many life science experiments involve using antibodies to assess outcomes, be it for immunohistochemistry, indirect immunofluorescence microscopy, western blotting, flow cytometry, ELISA, immunoprecipitation, or many others. Antibodies are so prevalent that you will almost certainly either use them yourself or will read papers where they are used. Therefore, I want to point out some extra things when considering experiments using these important tools.

This is very important: no antibody preparation is perfect. Antibodies work by recognising their specific antigen; however, they also always have low affinity for other targets i.e. there will be non-specific binding. Indeed, this is true for most reagents, specificity goes down as concentration increases. How much of the signal is real from your experiment and how much of it is due to non-specific binding is determined by the relative affinity of the antibodies to their different targets, the target antigen's abundance relative to the non-specific targets, the concentration at which the antibodies are used and the stringency of the wash buffers. That's quite a list of variables, and that means that you should do a series of pilot experiments to determine the optimal conditions for your assay to maximise this *signal to noise* ratio. But how can you know what is true signal and what is noise? Controls, of course!

In terms of positive controls for antibodies-based experiments, you will need a sample that definitely contains your protein of interest, ideally prepared in a similar way to your test sample. Examples might include cells, protein extracts or tissue from samples where previous studies have robustly characterised your protein. If these samples are unavailable, then a purified protein might be appropriate for some techniques. Processing these positive controls samples with your antibodies will show you what signal you could expect to get when your assay conditions are correct.

Antibody negative controls

- Knockdown/knockout tissue (**best**)
- Cross-species (if non-conserved)
- Different tissue or cell type
- Pre-immune serum
- Antigen depleted
- Isotype-matched non-immune

Negative controls for antibodies-based experiments are always required. The gold-standard negative control is a sample where everything else is the same, but the target antigen has been depleted in some way. This could be tissue from knockout plants, animals or patients, or conditions where siRNA/shRNA has been used to deplete the mRNA. When you first publish using newly-made antibodies, most reviewers will require data of this type to be convinced that you have made a good reagent and have optimised its use appropriately. Following the initial publication, the burden of proof is somewhat reduced; however, you should still use the strongest possible evidence you can.

It may not always be possible to obtain the genetic knockout for the technique you are performing, so what should you do then? There are a few options, and the choice depends on what type of experiment you are performing and whether the antibodies are monoclonal or polyclonal.

First, you could look at the cross-species conservation of the antigen used to make the antibodies, and then probe a sample from a species where there is not conservation. For example, for antibodies raised against a peptide sequence that is present in the human form of a protein but which is not conserved in mouse, then using mouse skin sections could be an option for a negative control. The rationale here is that although you know that your target isn't conserved, the non-specific targets of your antibodies are likely to be present. Therefore, the *noise* will stay roughly the same, but you should lose the real signal in the other organism.

Second, you could probe a different tissue or cell type from the same target species, but which you know doesn't express your protein of interest. This option is not as good as knockout because the signal-to-noise ratio depends on the local abundance of the non-specific targets, which will be different between the target tissue and the control tissue. However, this approach will give an indication of the general non-specificity of your antibodies at the concentration used.

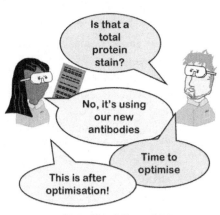

No antibody is perfect.
It is crucial to include appropriate positive and negative controls in all experiments.

Third, if you are using polyclonal antibodies, you might have access to the pre-immune serum. This is a sample of the serum taken before the animal was inoculated with the target antigen. The pre-immune serum will not contain antibodies against your target antigen, but will contain any antibodies that the animal has made against its environment. The pre-immune serum should give you the same *background* noise as your final antibodies and the difference in signal between the pre-immune and final serum should be due to reactivity to your target antigen. However, there is an important caveat: the non-specificity of your new antibodies is not just to the background, the new antibodies will also recognise other non-specific targets.

Fourth, if you have access to the antigen used to make the antibodies (e.g. a peptide or purified protein), you can use this to deplete your polyclonal antibody of the antibodies that recognise the target, to leave behind only the environmental antibodies. This should give the same signal as the pre-immune serum. This is less useful for monoclonal antibodies, as the depleted sample will remove all the primary antibodies.

Fifth on the list and the least stringent but also the easiest and most common is using an *isotype-matched* control. These

Big Tip
Your antibodies data sheet will tell you the isotype: IgG, IgM etc. Make sure to pick your isotype control and secondary antibodies to match.

controls are antibodies from a non-immunised animal. Any signal obtained using an isotype control is due to generic binding of *any* antibodies of the class (isotype) you are working with. Isotype controls are cheap and easy to do, so you should always include them in addition to other available controls but be aware that they are not the best option.

All the negative control options described above have caveats; you can still get a signal which isn't due to your target protein even after including these controls. This means that you will probably need to consider a combination of options.

Indirect approaches, where secondary antibodies conjugated to an enzyme or fluorophore are used to detect where the primary antibodies have bound, require additional controls to identify non-specificity of binding of the secondary antibodies. Always include a no-primary antibodies control, often referred to as *secondary only*. If using multiple primaries and multiple fluorophores, you will also need to confirm there are no cross-reactivity and no bleed-through of signal in your analysis platform. To do this, you will need to process samples individually as well as together.

Big Tip
You need controls both for the primary and for the secondary antibodies.

Sounds like a lot. You won't always need all of them, but your data can only ever be as good as your controls. It is better to include extra controls that you might not need than realise that you must repeat the entire experiment because you forgot to control for something. Don't be surprised if you decide you need more controls than samples in your experiment.

Controls to rule out alternative interpretation(s)

Remember how I commented about designing out third or confounding variables? Well, this is it. Time to think critically about the results the experiment will yield and try to identify every other potential reason why you could get the same outcome that would not be due to your experimental question. We're back to common sense and understanding of biology again! As usual, I recommend writing these down, as having a list in front of you can help you think how you might control for each one.

Big Tip
Try to control for everything that could influence the interpretation of your data.

Using stratification to control for population differences

In the next section, we will look at assigning groups, but I want to lay the groundwork here. If you are studying populations in their real-world context, then everything about that context could be a confounding variable. If there are environmental variables such as soil type, climate or geographic location, or things about the participants like age, gender, height, weight etc., then you can lessen the influence of these confounding variables by first separating your population based on those criteria before assigning them to experimental groups. We call this *stratification*.

The more variables you want to stratify for, the larger the population you will need.

Therefore, you will need to decide which confounding variables matter the most and prioritise stratification based on those most likely to have the largest influence on the outcome variables. When it comes to writing up your project, don't forget to include details about stratification in your methods section.

Example: Stratification

If you were designing a study to determine whether training at altitude has an impact on VO_2 max, you would go into the experiment thinking that the age range, gender and BMI of the participants are clear confounding variables, while level of underlying fitness, exercise frequency, and altitude at which the participant normally lives are all likely to be relevant. One option is to randomly assign participants to either the altitude training vs non-altitude training groups, and then account for demographic differences in the analysis. This approach would allow you to determine the effect of these variables on the outcome. However, if your question was narrower, you would obtain an easier to interpret and clearer message by stratifying the groups first so that they are roughly balanced in terms of these features. Both the fully randomised and the stratified first then randomised approaches can work in different context, what would be best depends on the study size, population breakdown, and s question. The best option is often a combination; stratify for largest confounders then account for the others through analysis.

Paired analysis

Measuring the same thing repeatedly within the same individual can be another effective strategy to remove the effect of confounding variables. For example, you could measure your outcome variables before and after treatment, or you could measure after treatment A and then again after treatment B. By designing your experiment in this way, you can control for the genetics, environment, lifestyle etc. In general, statistical methods for *paired analyses* are stronger than the equivalent unpaired design. This means you also are likely to need fewer samples to reach your desired confidence level (*significance*).

Paired set ups sound great but there are dangers that mean that you may not be able to use them. The order in which you deliver your manipulations can make a difference, or the effects of first treatment could remain active and influence the outcomes of the second treatment. It might be possible to test for these potential problems (a pilot study could do this) and thereby confirm they are not an issue or you could control for these order effects by randomisation; half your study population would get treatment A first, half treatment B, then reverse. There are also potential ethical restrictions that may preclude paired analyses; for example repeatedly performing procedures on the same animal is usually not allowed without clear justification.

Controlling for order and location effect

Order and *location* effects are issues to consider even when not doing paired analyses. In wet lab work or any procedural work, the sample that is processed or treated first may yield different results to the one processed last with those differences due to the order of processing rather than whatever manipulation was performed. In terms of location, the cages that are located closest to the door of an animal room might lead to more stressed animals, or the samples in the outside wells of a multi-well plate might give a different result than those in the middle due to different rates of evaporation (this is a real and commonly overlooked problem).

These potential issues are easy to control for by just *randomising* the location and treatment order both within an individual experiment and then again between different experiments. Because it is so easy to fix, there is no excuse not to protect your data against these potential confounders!

Controlling for bias

Every time a human is involved either as an experimenter or as a participant, you should consider how you can reduce the potential for bias. What can be done so that the people involved don't unconsciously influence the result? An obvious answer is that you should use *blinding* wherever possible. Actively look for ways in which you can prevent researchers and participants from knowing details which could influence their behaviour.

Be aware that in vitro lab work could also have a potential for human bias. Any time when you are using a subjective measurement system, such as the scoring of images, you can blind the scorers to which group they are scoring. You could consider investigator triangulation; multiple different investigators could conduct the same experiment to determine whether they obtain the same results, reducing the impact of any individual bias.

Don't forget that objective outcome measures can be subject to bias too; for example, a researcher might unconsciously take more care over the sample preparation from the treatment group rather than the negative control group. In all your research, you do not want your results to show what you hoped they would; you want your results to reveal the scientific truth irrespective of what that result is!

Use blinding to remove conscious or unconscious
sources of subjectivity from your data set.

Controls to calibrate the system

A fourth use of controls is to provide an internal measurement to compare your findings against. The simplest of these are the positive controls and experimental negative controls, such as empty wells in an ELISA, or untreated cells in a flow cytometry experiment. From these negatives, the background signal can be obtained and then subtracted from all your test samples.

Calibration can be used to convert experimental measurements into real-world relevant numbers. For example, a standard curve can be generated using known concentrations of your target molecule, and then by comparing outcome measurement against that curve one could convert an absorbance measurement into a concentration. If this is an option, you should take it. It is always easier to grasp the biological importance of a finding if it is presented in useful units rather than as an arbitrary scale.

A different form of calibration is internal normalisation. For example, rather than calculating the absolute abundance of your target mRNA, you might instead compare its level relative to some other genes, or you might use the level of a control protein or total protein abundance to normalise your samples and allow relative expression of the target protein to be compared. These internal calibrators are included to control for differences in pipetting or inaccuracies of your concentration measurements. In publications, you will see these referred to as *reference transcripts* (for RT-qPCR), *loading controls*, or in the older literature, *housekeeping genes* although this term is not recommended. What you should use as your reference points in a specific experiment depends on the experimental set-up and you will need a pilot experiment to identify which internal reference points are not affected by the assay conditions.

As with all other parts of your experimental design, you will need to be able to justify your decisions for normalisation before you attempt to make a broader interpretation. The normalisation will directly influence the data interpretation. As such, tou should always state any internal normalisation approach in your methods and they should be clear on your graph or explicitly mentioned in your figure legends.

Big Tip

Remember, you will need to be able to justify the choice of reference gene, transcript or protein. Part of your pilot studies should involve determining the best calibrator for your system and question.

Final comment on controls

Whenever a student is struggling with an experiment either in terms of getting it to work or in interpreting the findings, the very first thing I ask to see is the results from all the controls.

The reviewers of your manuscripts, paper or thesis will only believe your results if your controls are appropriate.

Spending time at the planning stage to make your experiments as tightly controlled as possible is worth the effort and will save time in the long run.

2.6 Independence and Randomisation

This next section covers two important questions that you must think about **every** time you set up a new experiment:

- Are my experimental units truly independent?
- How can I remove sources of bias?

What is my experimental unit, are they truly independent?

When you are analysing your data, you will perform a statistical test to determine how confident you can be that any observations you have made are real. Each number that goes into this analysis comes from one *experimental unit*. But what is that unit? Is it the measurements from one cell, one plate, one animal, one forest? The answer depends on how the populations are interconnected connected. If they are linked, then you can't consider them as separate data point. Each experimental unit must be truly *independent* of the others in the analysis group for it to count.

> **Big Tip**
>
> The experiment "N" is the number of *independent* experimental units.

Sometimes what is and is not independent is clear and unambiguous; cells in the same dish, and people from the same family are clearly related to one another and so you almost always cannot consider them as a being independent. However, it can be more complicated; is it appropriate to consider cells in the same incubator or from the same parental flask as being independent? Are study participants that have been recruited from the same hospital, school or even the same geographic region different enough for you to consider them as being unrelated? Ultimately the answer depends on what your core question is and what aspects of the shared environment could contribute to your experimental outcomes.

An important point here is that a statistician cannot answer this for you. It is you, as the person conducting the experiment, who must decide what level of independence you need by using a combination of your understanding of the biology being investigated and by using common sense. Because this is a decision, the choice you made should be explicitly clear in your write up and you should also be able to defend a decision against criticism.

Definition: pseudo-replicates

Pseudo-replicates are when you consider samples to be independent when they are actually related to one another.

Common Reasons for Pseudoreplication problems

If you perform statistical analyses on *pseudo-replicates* then you will over estimate your confidence in your results. A reviewer would likely reject a manuscript in this situation as your conclusions would not be supported by the evidence. Therefore, I want to point out some common things to watch for that can lead to problems.

- **Common environments**

    People/animals/cells that live in the same house, incubator, class, school, city, forest will share key aspects of their environment. If the environment influences the experimental outcome, then you cannot consider subjects from that common environment as being different from one another. Simple really.

- **Shared enclosure**

    Experimental mice/rats/rabbits etc. housed in the same cage **are not** independent of each other. Social animals respond to the groups they are in if you stress one animal in the cage then the others in the cage will also respond to that stress. This usually means that animals housed in the same cage must be treated as *technical replicates*.

- **Relatedness**.

    Parents, their offspring and siblings are obviously related. If one family member responds in one way, then the others will likely respond similarly due to their shared genetics or shared upbringing, culture etc. In lab work, the same holds true; all the flasks of cells derived from the same donor will be more similar to one another than those from a different donor. Therefore, especially if you are working with primary cells, then independence is usually established at the donor rather than at the flask level. Similarly, if you are making something like an artificial substrate or recombinant protein then independence might have to be at the level of the batch.

squeak

squeak

Mice, rats, rabbits are social animals, if you treat one animal in a cage, the animals that are housed in the same enclosure will also respond.
In the experiment above, the experimental unit would be the cage.
You would have N=3 for your statistics.
The 5 animals per cage would be regarded as technical repeats.

Biological versus Technical Replicates

Often scientists talk about *biological replicates* and *technical replicates*. Biological replicates refer to your independent samples; usually, this comes from how many times you did the whole experiment or the number of independent patients or animals used. Technical replicates, in contrast, are individual measurements you make from each biological experiment, e.g. the biological repeat might be the dish of cells whereas the technical repeats are the measurement from each cell within that dish.

In your final analysis, the technical repeats are conflated into a single number for each biological sample. If you had 1, 3 or 50 wells getting the same treatment per each experiment, you would still only ever be able use one value for the whole plate in your statistics. However, this doesn't mean technical replicates are a waste of time; to the contrary, they play an important role. By making multiple measurements of the same thing in the same experiment you can increase the accuracy of the final value that is entered as the experimental result for statistical purposes.

> **Big Tip**
>
> Increasing the number of technical repeats increases the accuracy of the measurement of each experimental unit
>
> **BUT**
>
> There comes a point where increasing the number of technical repeats add no additional value.

As you can imagine, there are diminishing returns with technical repeats, you only need to do enough technical repeats to obtain a reliable experimental value. After that point, you don't gain any further benefit by doing more. Use pilot experiments to identify how many technical repeats are needed. This is best explained with an example:

Example: Biological vs Technical repeats

If you are performing a scratch wound assay comparing two treatment groups, you might set up the experiment in a 24-well plate where 12 wells get treatment A and 12 wells treatment B. You scratch each well and make three measurements along the length of the scratch at the beginning and end of your assay i.e. 3x12 = 36 total measurements per treatment group. But is your sample size, your *experimental* N, 36, 12 or 1?

In this case, the answer is **1**. As far as your statistical test and the graph that you make are concerned, this experiment will give you one value for treatment A, and one value for treatment B. The 36 measurements you've taken per treatment are classed *technical* replicates. If the drugs/cells are expensive, or it takes a long time to analyze each image, you might have got almost as robust data using fewer measurements per experiment perhaps using three wells per treatment instead of 12. Use a pilot experiment to minimise the wasted time and resources.

Performing measurements through time?

You might have an option of using the same measurement method on the same subject but at different time points to obtain repeat measurements. Unsurprisingly, almost always these measurements could not be classed as "independent" but rather should be considered as technical repeats and then combined to give a single value per participant. The only way that you could claim independence would be if you could be sure that the first treatment did not affect the second treatment outcome.

We mentioned *paired analyses* previously; measuring someone before and after treatment. These set ups have some added statistical power but it is essential for you consider if the first measurement could affect the second. In terms of a biological response, this could simply involve ensuring that enough time has passed between measurements. However, in human experiments you have an additional consideration, psychological or psychosomatic effects can and will persist much longer than biological effects, and these effects may influence the subsequent measurements. Randomising treatment order and careful assignment to different groups is extremely important for this type of paired study, so let's deal with that next.

Randomisation and Group Assignment

Once you have identified what your independent samples are, you next need to think about how to distribute participants, samples, cells and treatments into the different analysis groups; i.e. who gets what in terms of treatments. This is where stratification can be beneficial to control for confounding variables. However, after you have performed any deliberate stratification, you thereafter use *randomisation* to avoid introducing any unconscious bias.

Mice respond Differently than Oranges:
A case-controlled study

Any questions?

When I said do a random controlled study, I didn't mean to study something completely random

Bench scientists beware! I see problems with lack of randomisation all the time in the lab, especially where the researcher hasn't thought about the implications of order effects. There is a tendency to perform the experiment in the same order every time; negative control first, positive control second, and treatment groups in position 3, 4, 5. While this set-up may be easier in terms of repetition and may feel like it is a well-controlled and logical thing to do, it automatically raises questions about whether the order in which you treated the samples or their relative location in the dish influences the outcome. Failing to account for order effects

Big Tip

You should be able defend your grouping and order decisions against the harshest critic.

could invalidate all your findings.

There is no excuse not to randomise. Note: do not rely manually introducing randomisation, as you will always unconsciously introduce patterns. There are lots of random number generators available online. This one is quite good and aimed at lab research – https://www.randomizer.org/

Non-equal group sizes (not ideal but may be ethically appropriate)

One final comment about group assignment; whenever possible, aim for a balanced design with the same number of samples per group. This will allow the use of stronger statistical tests in the final analysis. However, if the treatment is harmful or if there are risks of serious side-effects, then it might be ethically more appropriate to have a smaller treatment group and a much larger control population. These sorts of studies are more complicated to design, so ask for advice from a statistician to make sure you plan everything appropriately.

You should be considering the ethics of your design at all stages of the process so, although I am a strong proponent of designing experiments for easy statistics, this is one area were making things a more complicated is the right thing to do.

Writing Up

Every step of your experimental design involves some decision making. In your write-up, you must make these decisions clear to your reader. How you have defined the experimental unit, how you have assigned groups and how you have randomised the assignment of participants to groups are decisions that directly impact the strength of the conclusions you can draw from your data. When it comes times to write up, make sure you are absolutely explicit about each of these things.

I recommend including this information within a dedicated subsection for data analysis and statistics in the materials and methods of the manuscript.

It's not enough to just say "3 independent experiments"
You must include the details of what you have defined as your
experimental units in your methods section.

2.7 Sample Size Calculations

Now that we've designed everything we're onto to final stage, working out how many samples are needed. No, the answer isn't always 3! There is no one-size-fits-all answer. Some people panic at this stage but you don't need to. Sample size calculations are quite straight forward if you take them step-by-step.

The numbers of experimental repeats required depends on four main things.

- **What type of experiment are you planning?** Qualitative, hypothesis independent or hypothesis testing? The latter requires specific numbers which must be calculated before you begin.
- **How confident do you need to be in your conclusions?** Or, what would be the consequence of a false positive or a false negative?
- **How large an effect or how strong association would be meaningful in this context?**
- **How much variability from sample to sample will you have?**

Qualitative and exploratory studies

Consider coverage and variability
Formal statistical tests not normally required

For *qualitative* data where you will not be generating numerical data or will not attempt to make generalisable statements about the population, then sample sizes are not based on statistics but rather upon *coverage*. Your goal is to reach *data saturation*, the point at which processing additional samples does not lead to any new themes being identified. Use published studies as a guide for what numbers are appropriate.

If you are planning an *exploratory* study, where the outcomes will indicate an effect but where the plan is that you will design further experiments to test the veracity of that effect, then you also do not require a formal sample size calculation. In these cases, the design process involved thinking about how you will interpret the results. Examine published work to find similar styles of experiment, and you likely will find that there are established norms for sample sizes. Use the publications as a guide, but always consider the specifics of **your** experiment in your decision. Of prime consideration is the variability you expect between the biological replicates within each group, and how many comparisons you intend to make. As these numbers increase then so too should your sample size.

The outcomes from exploratory studies will provide the data that you use to design your formal hypothesis-testing experiments. Much like pilot studies, exploratory studies can provide two estimates that we need later: i) the size of the difference between the comparison groups or the strength of the association, and ii) a measurement of the spread of the data, i.e. how much between sample variability you can expect. Those two data points go forward into the formal calculations for your hypothesis test.

Confirmatory Hypothesis testing

Sample size calculations are required before experiments can begin

It is bad science to keep on repeating experiments until you reach the statistical significance level that you are happy with! Remember we are looking for scientific truth, not just the answer we want. Importantly: if you need to apply for ethical approval for your project then you will need sample size calculations in the ethics application. Similarly, if you apply for a grant to fund your work, you are likely to need *power analyses* there too and even if not explicitly required, it is a good idea to include..

To be able to calculate the sample size required, you will need to have an idea of

Big Tip
Sketching the graph that your experimental data will generate will make it easier to identify the statistical test required.

which statistical test you intend to use, and you will need to make some assumptions about the data. In Chapter 4 we will go into detail about how to use different stats tests. In the planning stage all we need to consider are some of the more common approaches. The good news is that if you have followed the

steps up to this point, sample size calculations are quick. I've broken it into four steps.

Step 1: Identify the variables and their classifications

You first need to classify the types of data you will be collecting. This is relatively easy once you have thought about what measurements you will make and what you will compare. Let's start with some definitions:

Types of variable

- **Independent variable**(s) (aka *factor variables*): either the part of the experiment that you will control, or the groupings you will use to break up your populations.
  - o *Factors* are variables that are manipulated by the experimenter. Each factor will have two or more levels, e.g. untreated and treated (two levels) or untreated, dose 1, dose 2 (three levels).

    An experiment can have more than one factor, e.g. you might examine different cell types as one factor and different drug applications as a second factor.
  - o *Treatments* describe combinations of factor levels, e.g. treatment A might be cell-type #1 with drug #1, treatment B could be cell-type #1 with drug #2, treatment C cell-type #2 with drug #1 etc.
- **Dependent variable**(s) (aka **outcome variables**): the data that you will measure, the things that you predict will change or correlate with the independent variables.
  - o **Primary outcome variable**. If your experiment plans to gather lots of data at once (i.e. test multiple hypotheses with one experiment), you come up against a *multiple-comparison* problem. As we discussed in section 2.4, you should consider what added value each

extra outcome will give you. The more outcomes, the more samples you will need. Often, when you consider the experiment in this way, you are able to identify one question that is much more important than the others. If this is the case then you can define this one question as the *primary outcome variable* while the others become secondary outcomes. Your experiment will rigorously test the primary outcome but at the same time gain exploratory information about other questions. You use the primary outcome variable to calculate the sample size. If you can't assign a primary variable then you will must use a statistical test which will account for the multiple outcome variables.

Classifying Variables

Once you have identified your variables, you next define what *class* of variable you have. Do this for **all** the independent and dependent variables.

- *Categorical, Binary* (aka *dichotomous*): only two values are possible. For independent variables, this could be group 1 vs group 2 or treated vs untreated, or, for outcome variables you could by a yes or no type question e.g. presence of metastasis, dead/alive.

- *Categorical, nominal*: same as binary but with more than two unrelated groups. e.g. control vs treatment 1 vs treatment 2. You can think of these as "labels" (binary scales are categorical but there are some different stats tests available when using only two groups).

- *Categorical, ordinal*: A special sort of categorical classification where the groups have a definitive *order* but where the groups are discrete and the distances between groups aren't linear, e.g. young/middle age/old, low/medium/high, A, B, C, D-type exam grades. Usually the classic five-point *Likert* survey responses "strongly agree", "agree" etc., fall into the ordinal classification.

- *Continuous; (interval)*: the data can be anything, e.g. age, height, force, temperature, speed, calories etc. We know not only the order but also the exact difference between the values. A continuous variable can be converted into categorical variables by defining cut-points (aka binning), but there must be good reasons for doing this.

Using these terms, it should now be possible to classify all the variables you have in your experiment. As usual, it will help if you write these classifications down. Example time:

Examples: Classifying variables.

Example 1: I'm planning a cell culture experiment where I want to know if a new drug treatment "X" changes the expression level of protein "Y" in one cell type. I have *factor* variables of *untreated, negative control-treated (vehicle only) and cells treated with drug X at a single dose.* These are *categorical* variables, and, as there is no order, they are defined as *nominal*. For the dependent variable, as I

---

intend to measure my outcome relative to a reference protein, those numbers could be anything, they will not be in discrete groups. Therefore, I have a *continuous outcome* variable.

Example 2: I'm planning an animal behavioural study where I want to determine whether or not my transgenic animals are more likely to develop cancer compared with animals without the transgene. Here I could either have *binary independent* variables; *transgenic* versus *non-transgenic*, or I could have *categorical independent* variables of *non-transgenic, heterozygous, homozygous* for the transgene (these could be classed as *ordinal;* the heterozygotes may have a partial phenotype). My outcome variable could be *binary*; cancer yes/no or it could be *continuous* if I planned to measure the number or volume of the tumours formed in each animal.

Step 2: Match the statistics test with your variables

This stage can be intimidating (at first) as all the stats tests have non-intuitive names! However, all you really need to do is match the stats test to the variables you have. At the back of the book (section 10.3) is a flow chart that will become more relevant for picking an appropriate stats test during data analysis but could be used here too. I've written this section in a slightly different way so that you can choose whichever works best for you.

Important: described below are the most common experimental set ups. If your experiment does not fall neatly into one of these options, then you should seek help from a statistician. It is always better to get help before you do the experiment than finding out that you made a mistake afterwards!

First, choose the subheading that describes the classification of your *independent* variables fall into, then, second, identify the type of the dependent variable(s) from the options:

Tests for *binary* independent variables:

- **Categorical outcome variable: *Chi-squared*:** Use when your independent variable and outcome variable are both categorical.
- **Continuous outcome variable: *T-tests*:** Use when you want to compare **two** groups and have a continuous outcome variable. There are three main types of T-test:
  - *Independent*: samples in study population A are unrelated to study population B, e.g. different wells/RNAs/tissue blocks/people
  - *Paired*: use when you make two measurements in the same sample, e.g. before and after treatment.
  - *1-sample T-tests*: use when you will normalise the control treatment to an experiment control or when you are comparing to a set value. For example, you might set the untreated sample to 1 or to 100% for each run of your experiment and then determine the treated sample level relative to this internal (calibration) control.

Tests for when you have *categorical (nominal)* independent variables:

- **Categorical outcome variable: Chi-squared**: Use when your independent variable and outcome variable are both categorical.
- **Continuous outcome variable:** *Analysis of variance (ANOVA):* ANOVAs can be thought of as "special" T-tests. They use the same underlying principle but ANOVAs also adjust for the extra comparison groups being considered. Use an ANOVA whenever you have more than two categories of independent variable. This group of tests are the most common for situations where you have "untreated", "control treatment" and "test treatment" setups. Again, there are different options:
  - *1-way ANOVA* – equivalent to an independent T-test. Each of your samples is distinct from one another, with each sample measured once.
  - *1-way repeated measures ANOVA* – equivalent to a paired T-test, you have measured the same individual experimental unit multiple times.
  - *Factorial ANOVA* – use this if you plan to analyse your data after splitting it up based on different factors, e.g. you might split the population into male and female and then determine if the treatment affects males and, separately, whether the treatment affects females. For two subgroups this may be called a *2-way ANOVA.*
  - *ANCOVA* – The added *C* stands for "co-variance". An ANCOVA is similar to a factorial ANOVA except that the additional level is a continuous variable rather than nominal/ordinal. This additional independent variable is termed a *covariate.*
- **Multiple outcome variables: MANOVA:** *multivariate*-ANOVA. If you are planning to have multiple outcome variables, e.g. are measuring more than one gene, protein or clinical symptom, then you need a test that accounts for the extra variable. MANOVA is one such option. If you define a primary outcome variable, you might not need this option. A MANOVA can be 1-way and 2-way like the ANOVA, and you can also do a MANCOVA, following all the same points as above but with now more than one outcome variable. However, if you got into this multi-faceted study territory you **definitely** should get a statistician's help!

Tests for when you have *categorical (ordinal)* independent variables:

- **Continuous outcome variable:** *Wilcoxon tests* **(for binary independent variables):** use for continuous data (essentially this is the ordinal equivalent of the T-test)
  - *Wilcoxon rank sum (U test)* use to compare group A with group B. (independent T-test equivalent).
  - *Wilcoxon sign rank test* used for paired samples (paired T-test equivalent).
  - *One sample Wilcoxon* use when you are comparing against one group

(as for one sample T-test).

- **Correlation with outcome variable:** *Spearman's Rho* or *Kendall's Tau* use these if you want to measure the correlation between an increase on your ordinal scale and increase/decrease in your continuous outcome variable.

Tests for when you have *continuous* independent variables

- **Binary outcome variable:** *Logistic Regression*: use when you have a binary outcome variable such as disease yes/no
- **Correlation with outcome variable:** *Pearson Correlation/Regression*: use when you are trying to determine if your dependent variable is correlated with your independent variable.
- **Correlation with outcome variable:** *Linear / non-linear regression*: use when you want to compare two lines of best fit, i.e. you have a combination of one continuous independent variable (e.g. time or age) and two or more categorical factors (treatment 1 and 2).
- **Binary outcome variable – time to event:** *Log-rank test*: Use this for "time to an event" type analysis. For example, you might want to determine whether patients died due to their disease (commonly used with Kaplan-Meir plots).

If you are well versed in statistics, you will have noticed that I have simplified all the options here to only discuss those that assume your data are *normally distributed* (*parametric data*). Once you have acquired the data, one of the first things you will do is test for normality and may need to change to the non-parametric version. Don't worry if you don't yet know what these things mean; we will discuss them further in Chapter 4.

Repetition in writing is bad.
Repetition in science is needed to let you believe your findings

Step 3: Plug your numbers into a power calculator

Not that you have identified what stats test you will need. The next step is to plug some numbers into a calculator to determine how many samples you need. This is known as a *power calculation*, and it is quite easy as there are free online tools available to help you.

I like; http://www.powerandsamplesize.com/

Most power calculations involve four numbers:

1. The ***effect size***: the size of the difference you want to be able to detect, e.g. the difference between populations, or the strength of the correlation. When defining the effect size consider the real-world importance of that effect. It is possible to design an experiment that would be able to confidently detect, for example, a 2% increase in cell migration rates; however, a difference that small might have no biological relevance. You should write the effect size into your hypothesis statement.

2. The ***variability***: the spread of the data within each of your study populations, e.g. the predicted *standard deviation*. These numbers will come from your

pilot or exploratory studies, or if you don't have those data then you might have to estimate these from published studies of similar experiments.

Big Tip

The effect size in your power calculation should be a biologically meaningful difference / association.

3. The ***P value (α)***: how confident you want to be that any differences you will observe is not due to sampling variability. That was a hard to read sentence. Let's try again. When you choose an α value, you are deciding what probability of false positive rate you're prepared to accept. This is known as the *type I error rate*. The final stats test you run after the experiment will return this type I error rate as a *P* value. The α value becomes your *critical P* value. If the *P* value returned from your stats test is below this critical number, then you will define the data as being *statistically significant*. This means that the probability you have a type I error is below your threshold. At the design stage, if you choose a critical threshold of 0.05*, this means that you are prepared to accept that in approximately 5% (1 out of 20 experiments) the differences or correlations you observe will be spurious. But is 0.05 appropriate (see box below)?

In your methods section, grant application or ethics power analysis you must report the α value you decided.

4. The ***Power (1-β)***: how confident you want to be that the *absence* of an observed difference is not due to sampling variability. Here you are deigning what false negative rate are you prepared to accept, termed the *type II error rate*.

Just like α level, the β level is a choice which you make before the experiment, and therefore, it is something that you must report as part of your methods.

Choosing confidence thresholds (critical P values and power)

Remember that it is you, the experimenter, that *chooses* the ***P* (α)** and **power (β)**.*P* values of 0.05 and power of 80% are very commonly used but it is important to acknowledge that it might be appropriate to design your experiment around different thresholds. For example: lower the critical *P* value and/or increase the power if you plan to exploit the findings in a clinical environment where it could be detrimental to patient outcomes if you based treatment decisions on a false-positive (type I error) or false-negative (type II error) result. Conversely, you might choose to accept higher α values for an early-stage study where a false positive would be less problematic.

As with all the rest of your experimental design, you should be able to justify your choice. An answer why you picked a P value of "that's what everyone does," is unlikely to impress an examiner!

Step 4: Final questions: sample availability, ethics and budget

Before you commit to starting the experiment, you need to decide if the sample numbers are viable. Four questions you need to answer:

1. **Can you actually acquire enough samples?** If you can't get the samples, you can't do the experiment! This can become a real problem in clinical research, e.g. a rare disease may not have enough people attending the local clinic to allow you to perform the experiment the way you planned. Knowing how many people are treated locally will let you know whether you will have access to the samples you need and will provide an indication of the timeframe needed.

2. **Is it practically possible to get enough samples?** You should consider whether you can physically do the experiment in the time available. Would processing 1000 samples take you a week, a month or a year to do? Is that length of time proportionate to the *value* the experimental result will give?

3. **Is it ethically appropriate to do the experiment as planned?** Can you justify this number of animals or donors when you factor in how important it is to answer the question?

4. **Can you afford it?** How much does each sample cost? Do you have the funding you need? Do a cost vs benefit analysis.

If the answer is "*no*" to any of these questions, then you will need to redesign aspects of your study (or apply for more funding). We all want to do the biggest and best experiments we can but sometimes it just isn't possible.

If you are forced to re-design, the first thing to consider is whether you actually need as many comparison groups as you initially thought. Removing one outcome variable or one comparison group can massively reduce the required sample size.

Big Tip Doing an experiment with too few samples to ever yield statistically robust data is a waste of time and effort, and can be ethically wrong.	An approach you might take is to do a small, exploratory study to narrow down the number of doses or the number of time points that you need to measure. You then would do your main study using fewer doses and fewer time points, and this would reduce the required sample size.

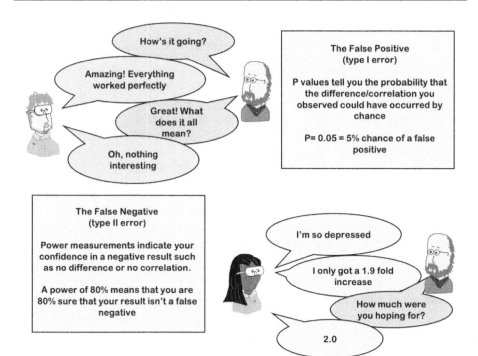

2.8   Experimental Design – Final Checks

You've got this far; you're almost good to go. Take a little time for three quick final checks before you kick into action.

1. Think about the ethics of your experiment again.

Seriously consider every aspect of the experiment again; the sample numbers, the amount of harm you will do to each participant, and what the data will mean in terms of adding value to the world's knowledge. If you have *any* concerns about the balance between value and harm, don't be afraid to step back and redesign aspects of your experiment until you are comfortable that the ends justify the means.

2. Think about all the ways the harshest critic of your experiment would look at your work and challenge your claims.

This "devil's advocate" test is important for being sure that your experimental data will satisfy reviewers and get published. Go through your list of controls, try to think objectively about confounders and alternative interpretations or ambiguity and if any concerns remain, then take a step back and try to fix them.

3. Talk over your experiment with someone else

A fresh set of eyes might be able to spot problems that you didn't. Now's the time to get the views of your lab head or senior postdoctoral researcher. Science is a team sport, so don't over be afraid to ask for help.

Still happy? Awesome! Get to work.

Chapter 3: Life in the Lab

3.1 Research Ethics
The rules for conducting research.
3.2 Tips for Lab life
New to a lab? This section will help you identify what you should do to fit in and be a good lab mate.
3.3 Time Management
Maximise efficiency to get ahead faster.
3.4 Lab Books
What you need to record and how to record it.
3.5 Data Management
How to store and organize digital data.

3.1   Research Ethics

There are a series of accepted research ethics rule to which all researchers **must** adhere to, irrespective of their location, background or discipline. These rules exist to ensure that research is conducted in a fair, transparent way that will stand the test of time and which will protect experimental subjects from unnecessary harm. Failure to follow these rules has dramatic negative consequences on a researcher's career. When you have lost the trust of the community, it is impossible to earn it back. Breaking the rules can also lead to you and your institution being sued or ending up in prison! The ethics rules are so important, it would be remiss not to have an overview in a book like this!

> **Big Tip**
> Don't ever cheat!
> Publishing fraudulent data will lead to *all* of your work being questioned and will effectively end your career.

We have already discussed some of the ethical considerations as they relate to experimental design. However, research ethics governs many aspects of your life as a researcher. Mostly these rules are about doing the *right* thing.

Fundamentals

Honesty

Let's start with the most obvious one; you should be honest throughout your work. This point doesn't only cover reporting the truth, but it also means being objective, transparent and displaying integrity in terms of the interpretation of the data. Deliberately stretching the truth in a grant application or using an inappropriate statistical technique to increase the significance of your research are equally as prohibited as falsifying data. Similarly, you cannot pretend not to know something or leave out data that would influence the interpretation of your results.

Responsible publishing

Central to the scientific ethos is that the reason to publish is to advance knowledge rather than solely as a means to advance your career. Accurate, complete reporting of research data is a key requirement to be ethically compliant. Remember that research that nobody knows about is the same as research that never happened. Therefore, you must publish your results irrespective of their implications.

There are two specific rules in relation to publishing that are important to be aware of *plagiarism* and *dual publication*.

Plagiarism is passing it off someone else's work as your own. This is probably the quickest way of getting thrown out of a University or failing a PhD, but surprisingly, it does still happen. Plagiarism happens in publishing too. Most journals and all universities use plagiarism checking software to flag papers with unreferenced sources. These software packages also check for similarity against all digitally available papers. Your reader or the journal editor **will** know if you have lifted a passage from some other body of work. The answer is simple; don't do it.

Write things in your own words and be sure to cite all your sources. Later in this book, we will talk about how to appropriately reference your source materials to ensure that you don't accidentally commit plagiarism offences.

Dual publication is a slightly different entity. You are **not allowed** to publish the same research in multiple publications. You are also not allowed **to submit** a manuscript to multiple journals at the same time. These rules include individual parts of a publication. Things like a single panel or image from a figure cannot be reproduced in another journal without directly citing the original publication. When you submit your manuscript to a journal, you will have to sign something to say that it has not been published elsewhere (there is no excuse for being unaware of this).

What to do if you suspect integrity issues with published work

When you read papers, you might spot mistakes or deliberate misrepresentation of the data. Sadly, this happens more often than you would hope. As a member of the scientific community, you shouldn't stand for it; post-publication peer review is a duty of everybody. What to do next can be somewhat delicate but really shouldn't be.

If you think it is a genuine error and is quite a small mistake, then usually the best option is to let the authors know directly (there will be a correspondence address on the article). They *should* then contact the journal and rectify the mistake. If the mistake hasn't affected the interpretation of the data, this involves publishing a corrigendum to the article and is not a major problem. Obviously, it would be better if it didn't have to happen, but a genuine mistake can happen and it is always better for it to be fixed.

If the error is more substantive or you suspect data fabrication (fraud), then contacting the author may be more uncomfortable. It really depends on the person but if they are prepared to try to cheat to get ahead then they probably aren't someone you want to interact with! If necessary, there are indirect and anonymous options to report fraud and also protect yourself. A good first step is to put a comment on www.pubpeer.com a post-publication peer-review site. You can do this anonymously and the authors will receive an email notification about the comment. The authors can respond on Pubpeer and can also rectify the problem at the journal through the right channels. Good scientists will respond positively to any comments on their work and will take steps to rectify any issues or provide the proof that the data is legitimate. Another option is to contact the editors of the journal and/or the ethics board of the University where the work originated, again you can do this anonymously to protect yourself.

Have a read of www.retractionwatch.com for other advice on reporting research integrity violations.

Carefulness

The research you will perform will have some value to society. Accurate results, irrespective of whether they support your hypothesis or not, will advance knowledge. But only if the data is sound. The ethical requirement to be diligent and take due care in data acquisition and recording is even more compelling when the subjects are exposed to harm. Poor quality data due to a lack of care is unforgivable.

Precision and carefulness not only relate to the reporting of data but also to your record keeping. It is an ethical violation to fail to keep good research records or maintain research data for a reasonable period. Lab books have multiple important roles in your research, and therefore, we will discuss those separately (section 3.4).

Openness

Scientific research should be shared. It is done for the benefit of humanity; therefore, a core tenet is openness. An important example of where you might fall foul of ethics rules is in your materials and methods section. The description of what you have done needs to be complete to the extent that another scientist could replicate your work exactly as you did it, importantly this includes all the data processing steps.

Many journals are moving toward a model where the openness ethos stretches further. Researchers are now required to submit all their primary (raw) data along with their manuscript or they must upload those data to an accessible online repository. Whether or not this is a requirement, you should still have all the data available and organised, ready to be shared upon request.

Openness also refers to the sharing of resources and data sets. Publishing your work indicates to the scientific community that any tools developed are available for others to use.

On a wider, more philosophical level, there is a move within the scientific community toward "open-access" publishing. This makes

Publishing your work in open access journals will make it accessible to more people.

published work accessible to anyone who wants to read it (sometimes after a time delay) and is a way of ensuring that research can reach the widest possible audience rather than being restricted to wealthy institutions. Major funding bodies around the world are beginning to make publishing in an open-access format a requirement for funding support. Even when it's not a specific requirement, you should take the open-access option if you can, and you should be willing to share your publications as far and wide as allowed by copyright laws. This isn't just ethically correct, it is also better for your for your work to be more widely read.

Respect for colleagues

At all stages of your research career, you are ethically obligated to treat your colleagues with professional respect. Many behaviours are specifically covered; let's look at the three big ones: authorship, confidentiality, and personal attacks.

> **Big Tip**
>
>
>
> Have open, frank discussions about authorship and author order early and often during the project

Authorship. In terms of ethics, putting someone onto the author list of a paper that shouldn't be there, or failing to include an author who deserves to be listed are both viewed as ethical violations. Authorship is a very murky field, what is expected differs between disciplines and between institutions, so the best advice is to have a clear discussions about authorship early on and then at frequent intervals as the project proceeds. Also, do not be surprised if author order changes as the project evolves and as different people contribute to the overall outcomes. It may also change between first submission and final acceptance. In section 6.3, we will go into a bit more detail about what constitutes authorship.

Confidentiality has personal and professional implications. The personal ones are obvious, you shouldn't share information about your subjects or colleagues. The professional ones include not be allowed to discuss with your colleagues the data from a paper or grant that you are reviewing, and you are not allowed to *use* data, ideas, or new methods that you learn about while reviewing a grant or a paper. Other relevant behaviours are respecting existing intellectual property rights and informing a collaborator of your intent to file a patent.

Personal attacks such as making derogatory comments in your review of an author's submission, sabotaging someone's work, using a racist epithet in the laboratory or promising a student a better grade in return for sexual favours are not only ethically wrong but are also illegal. If you experience anything like this, please report it so that the offending individual can be removed from the scientific community, to protect yourself and others from their behaviour

There are definitive rules about what the contribution required to be credited as an author

Legality

Stealing supplies, books, or data, exposing students and staff to biological risks in violation of your institution's biosafety rules, and breaking copyright law by making unauthorised copies of data, papers, or computer programs are all examples of illegal activities. Don't do it.

Perhaps less obvious is that deliberately overestimating the clinical significance of a new drug to obtain economic benefits or owning over $10,000 or equivalent in stock in a company that sponsors your research and not disclosing this financial interest, are also explicitly prohibited. Financial support disclosures must be included in manuscripts and presentations so that the readers, reviewers and editors can make their own opinion of any potential conflicts of interest.

Protecting human subjects

Ethics rules for research involving human subjects have well-defined rules that have been agreed within the international science community (the *Declaration of Helsinki*). The declaration firmly establishes that the research subjects' welfare takes precedence over the interests of science and society, and that ethical considerations take precedence over laws and regulations. The fundamental principle is to maximise respect for the individuals involved. This includes establishing the participants' right to make informed decisions about their participation in research at the beginning, throughout the research, and even after the research is complete (consent can be withdrawn at any time). Specific consideration must be given to providing safeguards for individuals who cannot protect themselves or who cannot give informed consent such as young people and those with cognitive difficulties.

When conducting human participant research, you have a legal and ethical responsibility to maintain the confidentiality of your subjects. You also have a social responsibility to perform non-discriminatory research. This includes protection of personal data in terms of storage and access (locked office or filing cabinets, password-protected data files). An additional requirement is that you must report any adverse event in a human research experiment.

In addition to the broad rules of the declaration, each country has ethics rules governing procedures for human research. Local committees or boards need to approve study plans before any research can take place and the approved plan must be adhered to throughout your study. Separate approval must be applied for if you wish to make any modifications. Make sure you are fully aware of all the local rules, that you have read and fully understand any ethics applications that are already in place, and that you understand the record-keeping systems that you are required to use.

Hippocratic Oath / Declaration of Geneva

Members of the medical community are sworn-in using a variation of the Hippocratic oath. I want to highlight some of the phrases* that apply equally to research involving humans as subjects:

- I will maintain the utmost respect for human life from its beginnings, even under threat, and I will not use my specialist knowledge contrary to the laws of humanity.
- The health of those in my care will be my first consideration.
- I will respect the secrets that are confided in me, even after a patient has died.
- I will not permit considerations of age, disease or disability, creed, ethnic origin, gender, nationality, political affiliation, race, sexual orientation, or social standing to intervene between my duty and my patient.

Phrasing from the Declaration of Geneva as amended by the World Medical Association, September 1994, and by the University of Liverpool's Faculty of Medicine, January 1995.

Protecting animal subjects

Animal research should only be performed where justified in terms of the added benefit gained relative to the harm caused. The experimental approaches should always be selected so that they do not cause unnecessary harm or suffering. You are also ethically obligated to minimise waste and thereby reduce the number of animals that will experience harm. Be aware that poor experimental design leading to too few animals being used for you to make a meaningful interpretation of an experiment could in many ways is worse than using more animals than required. Too few animals will mean that *every* animal in the study is wasted. Always get advice from a statistician when designing animal work.

Ethics rules protect those that are unable to protect themselves

As with human work, you must follow local rules in terms of what licences and approvals are required to be able to do your animal research. Different countries have distinct regulations in terms of which types of animals are covered, but most consider all mammals, birds and fish beyond a certain developmental stage, as well as cephalopods such as octopus and squid as requiring specific approvals. Once you have approval, you must work within those restrictions, maintain records and report as required.

Read more:

https://www.niehs.nih.gov/research/resources/bioethics/whatis/index.cfm

3.2   Tips for Lab Life

Some tips for working in a lab environment, especially for those new to research.

General tips

Ask Questions

As a scientist, your job is to ask questions, and that overarching premise holds for general work in the lab too. If you are ever unsure about something, ask! The people you are working with will not expect you to know everything, or even to remember everything from the first time they tell you something. For sure, you are not going to know where things are stored, how to dispose of waste, how to use machines etc. If you are worried that you are bothering people, then ask different people each time. Many of your early questions will be able to be answered by any experienced person working in the lab.

For every point, there is a counterpoint. As someone who has trained lots of new researchers, the only time I get frustrated with questions is when I think my trainee is asking me *instead* of thinking for themselves. It's better to seek confirmation rather than direction; "I think I should be doing …., is that correct?" rather than "What is it that I should be doing?"

Try to avoid questions where the answer is "have you tried looking that up"

Set aside time to read

You'll find every aspect of being a researcher easier if you read more. Get into a habit of reading a paper or some other information relevant to your work whenever you have a spare moment. Better still, make reading part of your daily routine; read something as you have your morning coffee or on the bus home, or whatever. I find reading a paper last thing before I go home works really well as I then process the information during my commute.

Stay up to date with new literature by setting up publication alerts on key search terms*. However, don't only read the stuff directly related to your work, give yourself a strong grounding by reading work in different areas. At the early stages of your career, you will want to build up a "tool kit" of techniques. Whenever you read a paper which uses an approach that you are unfamiliar with, take some time to read about the strengths and weaknesses of that experimental technique. Help it sink in by thinking about how you could use that approach to ask a question relevant to your work.

As you know that others will have also set up these keyword searches, make sure you include the common search terms of your field in any paper you write to aid discoverability.

Write Regularly

Get into the habit of writing often. Join or start a writing group. Writing is something that needs practice. Writing frequently will not only reduce the amount of work needed at the end of a project but will also improve your general writing ability, and therefore improve efficiency in the future. Methods sections, review articles or literature reviews, and public engagement pieces like blog posts are all valuable things that you can write before you have any experimental data.

Becoming part of the Lab

Respect your lab-mates, the equipment, and the lab

Regardless of how much you want to work on your own, you will be dependent on others for many aspects of your work. How you interact with people will determine how they respond to you. Be understanding of their needs. Choose to be on time for meetings and appointments and let people know if you are going to be late. Be courteous in all your interactions, say please and thank you.

Respect extends beyond direct interactions; it also refers to everything about the work environment. Tidying up after yourself, and performing lab chores like stocking, cleaning, and making up communal buffers are all part of work in a research lab irrespective of your career stage. If you aren't actively helping the team, then you are hurting the lab's progress.

Align your expectations with your supervisor's

Try to build a good relationship with your supervisor. Everyone's personality is different, so it's hard to give generalised advice. However, the answer often lies in *aligning expectations*. Understand what your supervisor(s) expect from you and appreciate what you can expect from them in terms of time and interaction.

My advice is initially to treat your supervisor formally and follow their lead in terms of relationship development. Start by calling them Dr of Prof and only change to first names once *they* have established that it is OK. Address your correspondence professionally. Even once you have established a relationship, remember that they are still your boss!

It is important for your mental health to have a life away from the lab. Set clear boundaries and don't feel that you must reply to emails at all times of day and night. Take holidays and don't try to work every minute of every day. However, the reverse is also true. Be realistic with your expectations of what you will get from your supervisory team. If you want feedback on a draft, don't expect them to drop everything to work on your manuscript. Give them time and tell them about deadlines. Remember that your supervisor has a life outside the lab too. They might answer your email at 11.59pm one day, but you shouldn't expect them to do that every time. The more you can do to make things easier for

Some expectations are easier to align to than others

your supervisor, the more time and effort they will be prepared to put into helping you.

A little personal note; working with my PhD, MSc, MRes, and undergraduate students is a highlight of my job. I get a lot from the interactions beyond the benefit to my lab's research. Irrespective of the student's academic abilities or how good they are at conducting experiments, it's the people and the interactions that I remember.

Become part of the team

You will be working not just with the supervisor but with the entire research group. You should do everything you can to become part of the team. As a new lab member, you are likely to be less busy than the experienced people, and you also are the person who is likely to need the most help. Therefore, foster some good feeling by pitching in whenever are wherever you can.

Big Tip
Research is a team sport. You aren't competing against your lab mates, you are working together toward the same shared goal.

Learn what your lab mates are doing, what their experiments involve and the approaches they are taking. Doing this will help you grow as a scientist, plan your experiments and understand papers that use those techniques. Learning what else is happening in the lab will also help you to identify who is the best person to ask specific questions.

Don't be surprised if you get assigned a lab chore such as making some specific buffers, cleaning or maintaining certain equipment, and, as you get more experienced, training new lab members in specific techniques. These are all normal ways to make sure the research group works at maximum efficiency. Being able to train new people is a skill that will be helpful for you in your future career.

Attend lab meetings, journal clubs and seminars

In addition to spending time in the lab acquiring data, there will be other activities associated with lab life and which will help your personal development. Most labs will have some form of weekly lab meeting that will be compulsory to attend. Formats vary from group to group, but these meetings are a good chance for everyone in the team to know what the others are doing and to catch up on lab-related news. The lab meetings primarily exist for the benefit of the researchers (i.e. you) rather than the lab head. Therefore, attending these meetings and contributing to the discussion should be one of your priorities, and you should schedule your week to make sure you can attend.

In addition to lab-related events, your department will likely run other opportunities to aid in your training. Seminars are great opportunities for you to learn about cutting-edge research in your field and to meet scholars from other institutions. Don't limit yourself to seminars that are directly related to your topic

Big Tip
Make the most of the seminars, journal clubs and other training opportunities put on by your department.

area, attend as many as you can from guest speakers and internal faculty, staff and students. The chances are that you could exploit the techniques being used to answer questions in your area. Even if that is not the case, you always benefit from knowing more.

Journal clubs are another great opportunity to learn. In a journal club you actively critique the selected paper by looking extra carefully at the experimental specifics, data presentation and interpretation. In so doing, you will become better at critiquing your own work, which is helpful when writing up your manuscripts. Again, make time to attend and if your department doesn't yet have a journal club, think about starting one.

Be nice to the technical team

If your research institute has technical staff then be nice to them! All the time. Their jobs are to make research happen more smoothly and easily for everyone working in the institution. They might do that by doing core things such as stocking, equipment maintenance, buffer preparation, and training, but that doesn't mean that they are at your beck and call. Remember you will need their help sometimes, and they will be much more willing to help the respectful person that is nice to them than the rude or overly demanding individual.

Before you Start Your Experiments

Be proactive about obtaining training

Whenever you start in a new institute, you will need to be trained in how to use specific pieces of equipment. Identify as early as possible what training you will require, who can provide that training, and when they can train you.

Big Tip

Plan to use complicated equipment soon after you have been trained in its use.

In general, get the training process started as soon as possible so that it doesn't hold you up, you will almost certainly be working to someone else's schedule and it might take longer than you would like. However, for complicated pieces of equipment such as high-end microscopes, there is no point being trained and then waiting for a long time before you use the equipment as you will forget what you have learned in the intervening time. Therefore, plan the timing of your training such that you can follow up by using the kit as soon as possible.

Wow. Pretty serious stuff there Conro

Mmm Not really

So why the suit?

I ticked the wrong box on my Risk Assessment

And the safety officer said I have to stick to it

Safety is a serious issue. Fill out your forms carefully and pay careful attention to handling and disposal requirements.

Complete your safety documents: Risk assessments and COSHH

You should always endeavor to know what risks are associated with your work, how chemicals should be handled, and how to dispose of them. Different institutions and countries have different laws, but fundamentally, the advice is the same. For any new chemical, record all the details you need. You should always know what the emergency procedures are for different chemicals or other substances. Pay attention to storage, protective clothing and equipment required, first aid and disposal procedures. Before you begin any work, make sure you know the evacuation procedures, first aid kit locations, and what to do in an emergency. Note that you have morale and a legal requirement to work safely. If you endanger others through your actions you are liable. Take safety seriously.

If you are performing genetic modification, radioactive, animal or human studies you will also need to know the specific regulations governing your work, the requirements for record-keeping and the different disposal routes for the contaminated samples or tissues. Falling foul of these regulations could cause revocation of licenses, and this would have a knock-on effect not just on you but also on other people in the group or institute.

Conducting Experiments

You don't have to invent everything! Make use of lab protocols and published work

Use the experience that is around you. Most of the experiments you do will be variations on something published before. You will need to modify the protocols so that they fit the question you are asking, but fundamentally, the core procedures are likely to be similar. A related point; others will have encountered the same problem as you. The answer to your problem is almost certainly on a forum or troubleshooting page somewhere. Don't beat yourself up reinventing something, search for a solution!

Big Tip

Make sure that all the reagents you need are available and all solutions are prepared before starting an experiment.

Read the protocol and prepare regents before you begin

Every person that ever trains you and every kit you purchase will tell you the same thing: *read the whole protocol before you begin*. One really important thing to identify is which reagents are required. Even when you buy a pre-made kit, you often need to supply some additional materials. Make sure you have all the chemicals in stock and that you prepare any complex solutions ahead of time.

Reading the protocol fully allows you to plan your day/week with approximate timings for what you will be doing at different stages and therefore prioritize your work to be as efficient as possible.

Understand what every step of your protocol is for

You not only need to know *what* to do but also you should know *why* you are doing each step. The established protocols (especially in premade kits) mean that simply squirting the right things into the right tube in the right order will almost always give you some data. That sounds great, but if you don't know what the steps are for then you may miss important pieces of information, things that could influence your experimental outcome or interpretation.

The biggest benefit for understanding the experiment will come if you need to troubleshoot something that isn't giving you the data you expected. If you know what is going on at each stage, you will be better able to interpret what has gone wrong and what changes to make. Understanding the protocol will also help you to identify what stages can be adapted to suit your question, which parts need to be precisely controlled to minimize experimental variability and which can be adjusted to suit your available timeframe.

Focus

As you work in the lab, you'll quickly get to a stage where things become routine and you actually don't need to think very hard about the techniques you are using. However, don't get too blasé; make sure you are fully switched on, and that there are no local distractions during the times that matter. Don't be afraid to tell people to leave you alone so you can concentrate and don't disturb people who are clearly "in the zone".

Whenever you are doing something entirely new, the best plan is to reduce or remove other experiments from your daily plan. This particularly true when it is a multi-step experiment spread over many hours. As a supervisor of a research team, I want my students and staff to generate as much data as possible. However, I would prefer you did one good experiment that worked, rather than try three where you made mistakes, wasted resources or which were poorly designed. Slow and steady will win the race.

Standardisation

Poor quality input material can only give you poor quality data. This is a really big and important concept to absorb early and then never forget. Always be prepared to stop an experiment if things aren't right. Be critical at every stage of every experiment to make sure that you aren't introducing an experimental variable that you don't want to.

The biggest problem in new researchers is a lack of standardisation of the simple things that they may not realise impact their experiment. Sometimes it is obvious; if you were doing a cell-based experiment and had twice as many cells per group in experiment two compared with experiment one, then you shouldn't be surprised if the outcome you are studying is different. But sometimes, it is less apparent. If the cells you were using had been passaged or fed on a different schedule between your different experimental repeats, then it might not matter that you have treated them identically after that, the biological difference could still be there.

The thinking that you did during your experimental design will hopefully have identified variables that you need to standardise. However, when it is a new experiment, you may not have been able to identify everything until you get into the lab and work with the real material. This is where the importance of carefully recording every aspect of your experiment really comes to light when you are trying to look back and work out what could explain strange findings. Simply, it is better to be careful and keep *everything* consistent.

Make sure reagents are re-ordered before they run out

Having to stop your experiments because someone has forgotten to reorder a key reagent is incredibly annoying. In practical terms, this means ordering replacements *before* things run out. This is especially true when you are using communal resources.

Always make sure there is enough left not only for your next experiment but also for others to complete their experiments.

Tell people when things get broken.

Things break, things get left out or go off, people make mistakes. This stuff happens, we all know it, we've all done it, don't worry about it.

It is really important that you don't just walk away if you find a problem,

Finding key resources missing when you need them is a massive source of frustration. Keep your lab mates and yourself happy by re-ordering reagents before they run out!

do *something* about it. If you don't know how to fix it, then tell someone who does. It's much more annoying and disruptive coming to use a piece of kit that has a problem which no one has done anything about than it being broken in the first place.

Dealing with Problems

Not all your experiments will work!

Despite all your careful planning and your well-designed protocol, things will still go wrong. Your positive and negative experimental controls will help you identify where things have gone wrong, and troubleshooting guides, local expertise and online forums will help you resolve the problem. But, sometimes, none of that helps. What then?

Everybody's experiments fail sometimes. The secret is not letting the set backs get your down!

Option one: swallow your pride and ask someone else to do it. Getting someone to help isn't a sign of failure; it's a sign that you are mature enough to realise that it is the best course for project success. It is a sign that you care more about the Science than your personal pride. Sometimes the fresh pair of eyes will spot a mistake, and you'll be able to move forward. Sometimes it will work for them for no apparent reason, they'll do everything exactly the same as you, but for some magical reason, the outcome will be better. That is frustrating, but at least you know the design was right, and you will be able to move forward. Most of the time, the second researcher will come up against the same problem as you.

Option two: give it a rest. You can end up too close to a problem to be able to see the solution. Focusing on a different aspect of your project for a while can help you see more clearly what has gone wrong. This can be difficult if the problem area is rate-limiting for the rest of your project; you may not have anything else you can work on. Discuss it with your supervisor and consider taking time to read or write up some other aspect of your work, or take a few days off, then come back refreshed and refocused.

On a personal note, I found it best to have multiple experimental streams on the go at once. I would try to have something with low complexity and high success rate that would constantly yield data as well as the more difficult but probably more exciting new thing. That way, my levels of frustration at encountering problems was always tempered by having something else that worked each week.

Try not to compare your progress against others

I know this is almost impossible, but everybody's project is different and will progress at different rates. There is no advantage in comparing yourself to others. When things aren't going well, think about what *you* can do to move forward. Your energies are much better focused on *your* project's next small step and how *you* can get beyond your current problem than worrying about how you stack up against your peers or colleagues.

3.3 Time Management

Planning your workdays

Work out how long the different steps of your experiments take and plan your day around those steps. At first, it is likely that you will be doing a limited number of hands-on experiments per day; however, as you progress you will need to arrange your day so that you can complete multiple parts of your project at once. Everything takes longer than you think at first, so plan in some wiggle room.

To work efficiently, you need to identify the steps that are going to take the longest and prioritise those steps; then fit the shorter tasks around the big stuff. Sound familiar? This is the same rate-limiting step concept as for project planning but now we are looking at parts of your workdays.

Almost all life science experiments involve an incubation step along the way where you are effectively just waiting for something to happen. The most efficient people are those who use the incubation periods of one experiment to advance some other aspect of their work. This could be writing up your methods or lab book, crunching some data, ordering reagents, or it could be running one of your other experiments. It takes time to get good at this. The first step is to think about experiments not as the total time required but rather as the *hands-on* time when you are physically doing something. In the example below the grey, boxes are waiting times when you could be doing something else.

	Experiment 1		Experiment 2	
9	**Prepare Resolving gel (10 mins)**			
	Gel Setting (30 mins)		**PCR set up (20 mins)**	
10	**Prepare Stacking gel. (10 mins)**			
	Gel Setting (30 mins)			
11	**Load gel (15 mins)**			
			PCR running	**Prepare agarose**
	Gel Running (60 mins)		(200 mins)	**gel (5 mins)**
12				
	Set up transfer (10 mins)			
13				
			Load agarose gel (5 mins)	
14	Transfer running	**Prepare blocking**	Gel running(30 mins)	
	(120 mins)	**solution (5 mins)**		
15			**Image gel (5 mins)**	
	Add blocking solution (2 mins)			
16	Block (30 mins)	**Prepare antibody**		
		solution (5 mins)		
	Add antibody solution (2 mins)			
17				

Maximising your working efficiency isn't just about an hour-by-hour plan to fill your day, it is also about keeping your eye on the whole project to make sure you are putting the effort into the right areas. When you are putting together short-term plans, you should do so with the project plan and priority list in front of you.

Plan time for data analysis and figure preparation.

Remember, an experiment isn't finished until it's been written up. Plan time into your week where you crunch the numbers and make draft figures. Pilot experiments should be analysed straight away, and confirmatory studies analysed once all the experimental repeats are collected. Analysing your data might reveal interesting findings. When it does, you will want to have the time to explore those findings further. If you wait until the end of a project before you do your data analysis, then you might not be able to exploit any new, surprising or exciting directions. More on data analysis shortly.

> **Big Tip**
>
> Your experiment isn't finished until the figure is made. Get into the habit of analyzing data and making a figure mockup before you finish for the day.

Plan time for reading and writing in each week

I know, I keep making this point! Ask any PhD student who has just finished what they wished they had done differently. 99% of the time they will say some variation of "I wish I had read and written more during my studies". The best way to make sure you do something is to set aside dedicated time as part of your routine.

Don't take on too much

As you grow in experience, you will be able to focus on more and more different things at once. However, you need to stay self-aware and don't try and more than you can manage. If you overstretch yourself, you may end up rushing and messing up experiments.

Work-life balance

One of the perks of our job is that we can work whatever hours we want (health and safety rules permitting). Your supervisor shouldn't be asking you to work hundreds of hours a week, but they will expect some level of professionalism. Don't be surprised if your supervisor expects you to keep to a schedule. Near the beginning of your time in the lab ask what working hours you are expected to keep, then stick to them (this is part of the aligning expectations point I mentioned earlier). If you are working in a lab as part of a degree program, then you should also tell your supervisor what other commitments you have including up-coming exams, so that they understand absences and modify their expectations.

Some people find the lack of formal schedule to be problematic, resulting in them losing focus and starting to drift into work later and then head home earlier. The opposite is also true, with no specific work hours you can end up working all the time. It is totally fine to treat science like any other job. Productivity in terms of outputs and the time you spend in the building are not directly related. It is how effectively you work rather than how long you are there that matters. Setting yourself a realistic work schedule is likely to help you focus on what needs to be completed within a certain time frame. It is important for your physical and mental health to take breaks. You will be far less productive if you are too tired or sick to work. Make sure you schedule time off and plan in physical activity.

Clearing your head by doing some physical activity might help form a fresh perspective to an ongoing problem. If the rest of your thinking is done while sat in front of a computer screen, then being disconnected might give a chance for uninterrupted thinking. I also strongly recommend having a mental distraction, something where you completely disconnect from thinking about science *at all* and are focusing on something else instead. This could be a sport, but equally could be something that requires mental focus such as art, games, dance etc.

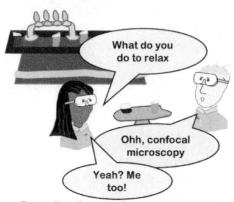

Regardless how much you love science, having interests outside the lab will help your physical and mental health

Email, social media and other distractions

Our phones are constantly buzzing with another update from emails coming in or the latest social media post. The ability to resist these distractions takes discipline. An effective strategy can be to dedicate specific times to deal with emails and restricting social media for breaks or when you have gone home. Most labs will have a "no phones" in lab rule anyway, so you might not have an option. Like every other rule, once you get used to it, it won't feel so difficult to abide by. Personally, I turn off notifications and close my email program whenever I am doing a task that requires serious concentration.

Note that there can be value to your science career to be active on social media, I'm not saying don't do it! Indeed, you will find a thriving community who want to share ideas or exciting data, and people who can answer your questions. Promoting your own work on social media can also help it gain exposure. As with everything else, it comes down to balance and priorities.

3.4 Lab Books

Why?

Lab books are a record of every aspect of every experiment that you do. They aren't optional, to keep a record of your methods and data is a requirement for any form of research in academia or industry settings.

> ## If it is not in the lab book, it didn't happen.

Lab books serve four important purposes:

- **They allow others to follow you.** Anyone else who ever wants to conduct the same or similar experiments will use your lab book as a resource where they can find out what works and what doesn't.

- **They provide evidence to support published work.** If there are ever any questions about the integrity or veracity of your work, your lab books will be scrutinised. Absence of proof will be damning. Failure to keep accurate records will bring all your work into question.

- **They can be used to prove first discovery.** If any of your work now or in the future has commercial potential, then your lab books will provide important evidence to demonstrate first discovery when it comes to filing a patent.

- **They provide a record for you.** Although this is an important role for your lab records, the three reasons above are more important. Your lab book is a home for you to record all the specific details about each experiment that you will need to write up your work. But, always remember that your lab book and data records are primarily *for other people*, so make sure they are legible and don't use shorthand, obscure acronyms or secret codes!

Conro's decision to use hieroglyphics for his lab book continued to cause problems

What should you record?

Basically everything! Record in your lab book *all* the details that pertain to your experiment, right down to the very smallest piece of information. For example, you should include not only the name of the antibodies used but also the dilutions, catalogue numbers, diluents, incubation times and temperatures, wash buffers and duration etc.

You should also insert any primary data such as affixing gels or blots directly into the book. For digital data, you should clearly record the file names

> **Big Tip**
>
> Lab books belong to the lab.
> Don't take them home.
> Make copies if you need information for your write up.

and locations of the raw data and processed data so that it is clear what is stored where. The digital data itself should also be clearly identifiable and usually best to include a short descriptor file along with the primary data.

Your lab head will likely have their own specific requirements for you to follow. But, the key message here is definitely that the more complete the records your records are, the better!

> ## You should not leave work until the lab book is up to date.

How?

Keeping a good lab book requires discipline. At first, it might be necessary to schedule time each day to complete your lab book. However, the best way is to write as you are working by having your book with you in the lab. I see a lot of new students keeping little notebooks that they then transcribe into their main lab book. While this understandable, it isn't ideal. It doubles your work and you are much more likely to forget. Whatever you do, don't let catching up your lab book build up.

Traditionally lab books were hardback, bound A4 or legal-sized books. Many labs still use these options, but there are now electronic versions available. What you ultimately need to use depends on what your lab head prefers. However, there are standards:

Paper Lab Books

Rules

- Number and date each page.
- Do not remove any pages.
- Do not leave large white spaces between experiments.
- You may need to have pages signed and verified for accuracy daily or weekly (especially for commercially sensitive experiments).
- The lab book should not leave the lab building (make copies of pages if you need to).

Tips

- Write the experimental hypothesis or objective on the first line of each experiment.
- Include the thinking and planning steps and notes on interpretation of data.
- Create an index at the front to help you find important experiments.
- Use the back pages for things like buffer recipes that you may need to refer to frequently.
- Once you have made the figure for an experiment, stick it in alongside the description of the raw data.

Digital lab books

There are now many options for lab book software available and they can be highly effective if used correctly. Some of these are paid-for services others are free, it is likely that your boss will tell you what they use. My lab switched to using digital lab books a couple of years ago, and I am quite a big fan, they've worked well for my students and staff.

The main advantage of the digital versions is that you keep all the information in a single location. This includes direct links to the primary digital data, to the processed data files, to the figures and, eventually, to the manuscripts. In addition, everything is clear, legible and searchable. The digital files benefit from an automatic date and time stamp and the meta-data associated with files means that these temporal data are unalterable. This is useful for any intellectual property claims. Digital lab books are accessible from everywhere (cloud-based) and can be shared easily with collaborators all over the world. Finally, I like that the entries are lockable and unmodifiable and are easy to validate and sign off where required.

Big Tip

Get into a routine of not leaving the building until you have finished writing up your lab book.

That list of pros sounds great; however, there are some drawbacks to digital records: firstly, the initial entries take a little longer to type up than writing in a physical book (although this is counterbalanced by being able to copy/paste parts of entries), this might also change as using styluses on laptops or tablets become more common. Secondly, most importantly, entering the lab book information requires access to a computer/tablet which you may not may not be allowed to use in the lab for health and safety reasons. We're back to a problem of compliance again. Being able to write as you go, really is a big advantage. If it is not in the lab book, it didn't happen.

Make sure you keep on top of writing-up your lab book. If necessary, schedule time each day to stay up to date.

Leaving the lab? Leave the lab book behind!

The lab books are not yours, they belong to the lab. They **must** stay with the lab (as must your data) when you leave. Make sure you hand them over, and your lab head knows where everything is stored. Ask for permission to take copies with you but don't be surprised if commercially sensitive entries are not allowed to be copied.

3.5 Data Management

The lifeblood of your research is the data you generate. Those data need to be stored in a secure, logical way. The better organised your data storage is, the easier it will be for you to write up, and the better it will be for your lab mates and boss!

File naming

Set up a hierarchal folder system so that different types of data each have a clear home. Do this *before* you begin collecting data so that you keep things organised throughout. Whatever you are saving should have obvious, interpretable names associated with them. Include the date in the name and record the file names in your lab book at the point where you describe the experiment. I also

Use a file naming system that will stand the test of time.
Descriptive file names with dates.
Record the file names in your lab book.

suggest that you add a file describing the experiment to the folder where the data are stored.

Avoid using names that will feature in lots of people's work. I've received hundreds of documents and folders labelled "results" or "report" that have come from students who haven't considered that everybody else also has results or writes reports!

Version control

Version control is something you should adopt early. Don't call things "new" or "final" or "draft", instead the first version saved is v0.1, and then this number rises to 0.2 etc. as you add to it and save new copies. Ultimately transition to v1.0 when the draft is ready to be shared. The increase to v1.0 could happen after you have had feedback from your supervisor or it could be the submission version.

Data storage

Back everything up! Multiple locations, physical storage and cloud-based storage. Schedule automated backups of your computer. Trust me, the pain of losing a hard drive is tough. Solid-state drives are better than traditional hard disks, but computers and especially removable drives get lost, stolen or corrupted, so make sure you back up and don't rely on single storage sites. Also, make sure you regularly remove your data from any communal computer such as the ones driving your microscopes, qPCR machines, plate readers, flow cytometers etc. Those hard drives get filled up quickly, and it is your fault if your data gets deleted before you transfer it to your personal storage site.

Back up non-digital files too. Make copies, keep them off-site.

Password protect and anonymise patient details and lock cupboards or offices where you store sensitive material.

Chapter 4: Processing Data

4.1 Data Analysis
General tips for all forms of data analysis.
4.2 Data Processing
Key concepts in data handling, including descriptive statistics and definitions of common terms.
4.3 How to deal with Outliers
What to do with the data points that just don't fit.
4.4 Inferential statistics
Testing for statistical significance, including choosing the appropriate test.
4.5 Writing about statistics
How to turn your numbers into sentences people will understand.

4.1 Data Analysis

Data analysis is a key step on your path to discovery!

The data are in! You've done the hard graft of designing, optimising and carrying out the experiments. Now you get to the truly exciting part of research, finding out the *answer* to the question. Was your hypothesis correct? Have you discovered something new and interesting? This is part is the reward for all the time and effort it took to get the experiment to work.

When to analyse

Data analysis is part of every experiment and it takes time. Make sure you set aside some dedicated periods for analysis into your Gantt chart and weekly plans. Doing this part of the experiment in the appropriate way is just as important as the actual data collection steps.

For your **pilot experiments**, you should analyse your data straight away. Some of the reasons for doing pilot experiments are to improve your experimental technique, to determine effect sizes and to identify the variables that require controls. For these types of experiment, the ideal time to analyse is literally on the same day that the data comes in. The reason is fairly obvious; you want to know what to do next or what changes to make. However, that take some thinking. So, let's introduce the concept of passive thought. As you become more invested in your project, you will find that your mind wanders to whatever problems you are currently experiencing while you are in non-work situations, e.g. as you are walking to work, or sitting on the bus, or wherever. Having data analysed before you go home not only rounds out the day but also gives you a chance to really chew over what your findings mean, to sleep on them and think through them carefully. This helps you to be ready to put a new plan into action as soon as you arrive back to the lab.

For **exploratory studies** you will most likely have large amounts of data and also multiple experimental repeats to process. The analyses for bigger experiments may be much more involved and time-consuming; if you are working to a deadline, make sure you don't leave all the analyses until the last minute! In terms of project planning, you will use the data from exploratory studies to plan a hypothesis-testing study. Therefore, again you will want the analysis completed ASAP.

There is a big difference once you reach the stage of doing a **confirmatory study**. We're suddenly into territory with strict rules. Here, the rules are that your data must remain *sealed* until you have acquired the data from all of the experimental repeats as determined by your power analysis. The only thing you should do until that point is a small amount of quality control checks. The sealed data concept is in place to remove a potential source of bias and this rule must be adhered to.

Data entry and organisation tips

The first step of data analysis is to organise the data in a way that will allow you to process it easily. Usually, this involves entering the values into a either spreadsheet-type program like Excel or directly into a statistical package such as SPSS, Minitab, or

GraphPad. For qualitative data, you will might use a thematic analysis package like NVivo, again the first step is data entry and organisation. The packages mentioned above are all paid-for and you may get a licence via your institution, but there are also lots of freeware available, don't feel like you must use one of the big commercial suppliers.

Big Tip
Save multiple versions of your data files so that you can step backward if necessary.

Increasingly people are using the "R" programming language for data analysis, this is a good option with lots of flexibility but has a learning curve. It is definitely worth learning if you plan to have a career in research. Whatever software system you have available, get to know the tools, the strengths and limitations of that software and find an alternative if you don't have everything you need.

If you have complex data with many variables and many entries per variable, then having a consistent and logical organisation is key to making the rest of your analysis easier. Some tips:

- **Always keep a copy of your raw, untouched data**
  You may need to start again if you mess up, or decide to re-analyse in a different way. The raw data may also be requested by journals, reviewers, or people assessing your patent applications. Always keep a folder where the original data is stored before any manipulation. Include a document that describes what the data are, how they were generated, and, importantly, how the files are labelled. A scan of your lab book entries could (should) be appropriate for this file, or you could get ahead on your writing by including a draft of the methods section.

- **Establish consistent, intuitive naming systems for your experimental variables**
  You will share your data with your lab head, the rest of your research group, and ultimately with the wider research community. Make sure to label everything in a way that is clear not just for you but also to others who will be working with the data after you. Make your labels as complete as possible and include a key or description tabs to define what everything refers to.

- **Set up an empty template for your data before you start the experiment**
  You will know before begin what the data will look like and how you intend to interrogate it i.e. what question(s) your experiment was designed to answer. Laying out your spreadsheet or data table in advance will save you hassle later and reduce the potential for making mistakes when copying or transposing data later. The fewer steps you need to take in the data processing aspects, the lower the likelihood for making a mistake.

Spreadsheet tips

- **Use one big file containing all experimental repeats rather than lots of small files**

I do this by putting each experimental repeat on a separate sheet, then include a *consolidated* or *processed data sheet* where I bring the experiments together. When the experiment has technical repeats as well as biological repeats, then the consolidated sheet need only contain the biological repeat outcomes. Remember that you can temporarily *hide* anything you are not working with, so although this one master sheet might look busy as you enter the data, you can change the appearance when it comes time to analyse. Having the data all in one place and linked together means that it is easy to step back and look for any outliers, spurious values and to identify and correct mistakes.

- **Freeze panes**

This tool is present in almost all spreadsheet-type software packages. The freeze panes tool allows you to keep a selection of your column or rows visible, such as your headings, as you scroll down or across the page. This is useful if you are entering the data from many experimental units or many variables, especially when you have hundreds or even thousands of entries!

- **Conditional formatting**

When you first look at your data you will want to get a feel for how things look and try and spot anything that is out of place. My favourite quick check tool for this, is *conditional formatting*. This allow you to select a range of cells of your data and automatically apply a colour scale to it (e.g. green for high values and red for low). This is just colour formatting; it doesn't change the data in any way, but the applied colour scheme can help you spot problems.

Ah, a lesser spotted outlier. Good find!

Conditional formatting can help you spot trends or problems in your data set

I use conditional formatting for lots of reasons. In my raw data, I apply it to columns of equivalent data to very quickly spot any mistakes in data entry. For example, if all my data was red/orange and I had one green cell, then I would know to look carefully at the green cell to see whether I had entered that data incorrectly. I also use conditional formatting to allow me to spot patterns with respect to the questions I plan to ask. Formatting the columns that I plan to compare will give me a quick indication of whether I will see differences between the populations.

4.2 Data processing

Now that you've got your data laid out all nice and neatly it's time to start analysing. What you do next depends on what sort of experiment you have performed.

Formal hypothesis testing: test your hypothesis!

If your experiment was designed to test a hypothesis, then you should have already broadly planned how the comparisons will work. This is the point when you will be happy that you took my earlier advice about simplifying your experimental design and asking one question rather than three. Now you have the data, all you will need to do is *normality* tests and select the appropriate statistical tests based on your actual data. Likely this test will be the same as you used in your power analysis but the spread of the data may mean that something different is required. We'll come back to these points in section 4.4.

People often state that you can prove anything with statistics, but at this point, you should be aware that the **only** question(s) that it is appropriate for you to report statistics upon are those that you designed the experiment to answer. You cannot change the question based on the data you obtained. The other interesting observations that are now apparent can be reported, but they will need separate confirmatory studies to be tested.

Exploratory studies: go exploring!

If your experiment has not been designed to answer a definitive question, then you will likely have data from many variables. These data sets can be intimidating to start with as it will hard to know where to begin your exploration, but, as with all science, it comes down to asking questions. The more you do of this sort of exploration, the more you'll find it fun.

> **Big Tip**
>
> In exploratory studies, analyze all aspects of your data, not only those that relate to your specific experimental question.

At this point, my standard approach is to graph *everything*! I start with the obvious stuff, the questions that I had before I started my experiment. But after that, I look at everything else. Most often, these extra steps won't reveal anything or, at least, anything interesting. However, there is always a thought in my head that I don't want to miss something, and every so often you'll come across a surprising observation that can explain what is going on within the population but which wasn't part of your original prediction. These surprising findings are why you did a hypothesis-independent or exploratory study in the first place!

A word of caution. Every time you ask an additional question of the same data set the probability of a false positive goes up. If you don't consider *all* of the questions you asked when you do your statistical tests, your P values will be artificially lower than they should be. See the box on *P-hacking* later in this chapter. This is why that after doing a hypothesis-independent experiment, the next step in a project is almost always to do a confirmatory, hypothesis-testing follow-up study with a narrow research question.

Descriptive statistics

A good place to start with data processing, irrespective of where you are ultimately headed, is to generate descriptive statistics of all your main variables. This type of stats, as they sounds, are simply summaries of your data. The descriptive stats are usually the values that you report in the results section. Note that these stats are different from the *inferential statistics* that you will use to make assertions about the confidence you have in the observed outcomes (we're not generating P values yet).

Graphs help you visualise your data. Try different ways of plotting your data, you might spot something that you'd otherwise miss

For each experimental variable, there are usually three relevant measurements: the *distribution*, the *central tendency* and the *dispersion*. In non-stats langue: the spread, the average and the error. When comparing variables, you will also care about the *effect-size*, the size of the difference between the groups accounting for data spread.

Every stats package will be able to generate these values with just a few clicks, but there are a few choices about which version of each measurement should be used, so the decision you need to make is which to use and when to sue them. To make that decision, we need a little bit of explanation.

Distribution (or frequency distribution)

The *distribution* is a description of the frequencies of each of the observations across the range of all the observed values. Quickly visualised by plotting the data as a histogram, dot plot or box and whisker graph. The distribution could be how many of your experimental units had each different measurement, or it might be appropriate to group the measurements into meaningful *bins*. For example, if you were plotting body mass index of a study population, it might be biologically relevant to group them into "underweight", "ideal", "overweight" or "obese" using standardised BMI cut-off points. The distribution plot would describe what proportion of the study population was in each group.

We need the distribution data to make decisions about how to treat the data, and which central tendency and dispersion measurements are appropriate. So, what are we looking for?

Normal Distribution (Gaussian distribution)

In most situations, if you collect samples indefinitely, the samples will ultimately adopt a consistent, hence *normal*, distribution, almost symmetrically spread either side of the centre point. This sort of distribution is commonly described as a *bell-curve*. The application of this premise is that even if observed distributions aren't normal from the sampling that was done, if you continued to take repeated samples and repeatedly observed the means, then the distribution of those sample means will converge towards normality. In practical terms this means that larger samples are able to detect smaller effect sizes or, to put it another way, will have greater confidence that the observed effect sizes are not due to sampling error. This principle is known as the *law of large numbers*.

This law is a core concept in statistics. Many statistical tests are predicated on this theorem holding true and that your data are normally distributed. However, it is important to be aware that not all the data you generate will fit this assumption; you can have *skew* or *kurtosis*, or could have a *multi-modal* distribution. As will discussed shortly, testing for normality is usually the first step in choosing a stats test.

In practical terms, identifying whether your data adopts a normal distribution makes a difference in terms of what inferential statistics test you can perform.

Skew and Kurtosis

Skewness is a measurement of lack of symmetry. You'll see skew most commonly when there are lower or upper limits to your measurements, e.g. when the mean is just above zero or near 100 %.

Kurtosis refers to the tails, the width of the distribution curve. High kurtosis means you have long tails or potentially outliers, and low kurtosis means you have short tails. When you do a normality test, the kurtosis value will be returned with the outcome. For normally distributed data kertosis=3. Values above three are described as *heavy-tailed* distributions, below three are *light-tailed*.

If you have skew or kurtosis, then your data are not normally distributed. This means that you might need to perform a *data transformation* before applying your statistical test (you guessed it, more on this shortly).

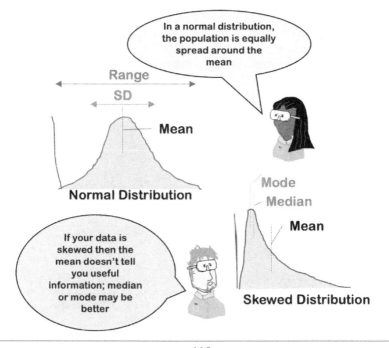

Central tendency (mean, median, mode)

The *central tendency* is an indication of where the middle of the data lies. There are three main options that each tell you different things about data distribution. All could be relevant depending on the data and on what question you are asking; *mean, median* and *mode*, or (rarely) none of these are useful at all!

- The *mean* is the simple average; add up all the values and then divide by the number of values. Undoubtably the mean is the most commonly reported value but don't blindly assume that it is the right metric for your study. Means are only useful when your data are symmetrical (normal or can have kurtosis) *and* when there are no large outliers in your population. For example, if you want to comment on the average income of your study population and the sample had 20 people who earned 20-50,000 and one who earned over 1,000,000, then the mean earnings would be around 90,000. While reporting this number would be accurate for the population, it doesn't describe the population well and therefore would likely not be useful. Better to use median or mode.

- The *median* is the central number of the population when the data are arranged in order from smallest to largest. If your data are symmetrically distributed, the mean and median are the same number. The median is most useful to when you have outliers or if your data are skewed. If the mean and median are quite different, then you almost certainly shouldn't use the mean. All stats packages calculate both in seconds, so no reason not to check how close they are.

- The *mode* is the most frequently occurring number. It is the least commonly used but it is the only really useful metric if you have *nominal* data, and could also be the best option if you are working with *ordinal* data. If two numbers occur with equal frequency, then you would have two modes and describe the population as *bimodal*, or *multi-modal* if you have lots of peaks on your distribution curve.

- **No central tendency**. Central tendencies might not tell you anything useful at all. If your data split into sub-populations (bi or multi-modal), the central point is unlikely to be valuable. Consider talking about the percentage within each sub-population instead.

Central tendencies (mean/median) usually don't provide useful information when your data are broken into sub-populations. Data that are spread like this are termed bimodal or multimodal distributions

The dispersion (range, standard deviation)

The *dispersion* is a number that describes the size of the spread of the data around the central tendency. In comparison, the distribution is more about the shape of that spread. There are three commonly used options: *standard deviation, range,* and *interquartile range.*

Big Tip

Usually the standard deviation is the most useful for the reader.

- **Standard deviation (SD)** is an estimate of the dispersion of the data relative to the mean (don't use SD with medians). As the mean is the most commonly used central tendency, the SD is also the most commonly used dispersion measurement. Your software will calculate your SD for you, however, you should know where it comes from. To generate the standard deviation, calculate the mean, and then subtract the mean from all the measured values from the population. Each of the resulting values is then squared, and the squares added together. The resulting *sum of the squares of the difference to the mean* is then divided by the number of samples minus one (N-1). The SD is the square root of that final number.
  Note: adding extra samples does not reduce your SD.
  Specific journals will define whether to use SD, s.d., sd or something else, check their "instructions for authors".

Standard deviations (SD)

In a normally distributed population, the standard deviation gives you some useful information:

- ~68% of the population fall within one SD of the mean.
- ~95% of the population fall within two SD of the mean.
- ~99% of the population fall within three SD of the mean.

- The **range.** This is simply the difference between the highest and lowest values in your dataset. Most often the range is used alongside medians or modes. When you use modes and medians, the highest and lowest values of your data set do not influence on the central tendency; therefore, the range provides additional important information about how the populations are spread. Note that this isn't an absolute rule, ranges can be relevant for means too, for describing how wide the data are spread or when you want to draw attention to some extreme values. For example, in a drug trial, the people who responded by the greatest and smallest amounts might be relevant to the decision about whether the drug was safe or worth prescribing.
- **Interquartile range (IQR),** sometimes known as the *midspread* is the difference between the 25th and 75th percentile (rather than the full range), i.e. it covers the middle 50% of your data set. Use the IQR in situations where the median is most relevant, i.e. where you have skewed data, or where there are small numbers of large outliers that don't necessarily reflect the general response. The IQR is conventionally

what the box in a box and whisker plot represents.

When you are writing about your data, mean and SD, or median and range/IQR go in parenthesis after the results statement and you should tell the reader what you are reporting (e.g. Sample 1 mean 10 SD 4). Measurement units go after the first number but not after the second (Sample B median 80 μm IQR 25 to 120).

Confidence measures derived from dispersions (SEM and CI)

The next two options are not measuring dispersion, they are confidence measures. You do see them reported in published work, and it is important to appreciate what they are telling you:

- **Standard error of the mean (SEM).** SEMs are calculated by dividing the SD by the square root of the sample size. Therefore, as the sample size increases, SEMs decrease. Although the SEM is directly related to SD, you use the two measures for different reasons. SDs describe the spread of the data around the *measured* mean whereas SEMs indicate the range within which the *true* mean of the population resides. The more samples you measure, the more confidently you can predict where the true mean lies.

Big Tip

Most journals and reviewers discourage the use of SEMs. Use SDs for spread, or CIs for confidence.

- **Confidence interval (CI).** SEMs on their own don't convey much useful information, but they can be used to generate a more useful measure; the *confidence interval*. The most commonly used is the 95% confidence interval; this is the range of values that you are 95% certain contains the true population mean. The 95% CI is calculated as the measured mean ± (1.96 x SEM). Note that the 1.96 in that calculation is the standardised score for 95% (you will need to look up a different value for different CIs). Of course, every stats program will calculate these for you.

SEMs and CIs are used to generate *P* values in stats tests, which you then report in your results. Therefore, reporting the SEM is of limited additional value and should only be used where you explicitly need to inform the reader about the *precision* of the study; i.e. how well the measured sample represents the entire population. Although, even if you do want to report the assay precision then the CI is the more valuable metric. You should not use SEM in situations where you are informing about population spread.

If you do decide to report either SEM or CI you must also report the experimental n alongside. Use phrasing such as (mean 10 μm ± 4 95%CI, n=5).

Effect size

When comparing two or more populations, the statistic you should care about is the *effect size*. From the name, you can easily guess that this is just the difference between the central tendencies of the study populations. However, this statistic also considers how spread the data are around the mean/median.

As usual, your statistics package will calculate effect sizes for you. But let's use an example so you can see how it combines dispersion and central tendencies. For a simple case comparing two populations that are normally distributed, you would calculate the effect size by determining the difference between the population means and then dividing that number by the *pooled* standard deviations of the compared groups: $\sqrt{((SD_1^2 + SD_2^2)/2)}$.

Example: Effect size

Comparing two populations for two outcomes.
Population 1: outcome A mean 14 SD 1, outcome B mean 10 SD 4
Population 2: outcome A mean 12 SD 1, outcome B mean 4 SD 4

Diff. in means	A = 14-12 = 2	B = 10-4 = 6
Pooled SD	A = $\sqrt{((1^2+1^2)/2)} = 1$	B = $\sqrt{((4^2+4^2)/2)} = 4$
Effect size	outcome A = 2/1 = 2	
	outcome B = 6/4 = 1.5	

So why do you need this effect size? Well the advantage is that the effect size is a number with no units, it is a straight ratio. This makes it possible to directly compare the effect of your intervention on lots of different outcome variables that use different units or that are in different orders of magnitude. This feature makes the effect size particularly useful if you want to say that a treatment had a bigger/smaller effect on one measurement than another.

4.3 How to Deal with Outliers

Outliers can appear in your data set for one of two reasons; they could be real data that is far from the rest of the population and represents true biological variability or they could be a spurious value that is the result of a mistake during the experiment.

> **Big Tip**
>
> "Outliers" are routinely defined as values more than three SDs away from the population mean.

How you deal with these potential outlier points depends on which of these categories you think they fall into, how confident they are part of either group and what the biological meaning would be if they were true values.

The decision between the options below are all subjective; **you must report what you have done** and you will need a clear rationale that you can defend.

There are five approaches to dealing with outliers:

1. Leave it in

Usually the best option. We are looking for biological truth and variability is part of that truth. The effect each individual outlier has on your data depends on the number of samples you have; the bigger your sample size, the lower the effect any one spurious value will have.

2. Try a transformation

If it is a true outlier, then you could try doing a transformation on the data rather than using the direct data to bring the spread back toward normality. Some options might be to *log transform* the data or to use percentiles rather than absolute values.

3. Drop the data

If the outlier **is definitely** an error, then remove the point from your data set, simple. However, if you are unsure if it is a mistake or not, then it *might* still be appropriate to drop the data. In 99% of situations, I **do not** recommend this, especially if you have a small data set of below ten experimental repeats. Biology has variability and you should be prepared to allow for that variability. You certainly should not drop the data without good reason and must report what you have done in your methods,

4. Use a threshold

For certain types of experiments, it may be appropriate to have a cut-off point above or below which the data is removed for some justifiable reason. Again, you must be explicit in your methods section as to what you have done and why. e.g. "Values more than 3SD from the mean were excluded from subsequent analyses".

5. Assign a new value

If the outlier seems like a mistake, it *might* be appropriate to impute a value to replace it. How to deal with missing data is a field unto itself; there are accepted techniques to infer what the value should have

> **Big Tip**
>
> You must describe *everything* that you do to your data in your write up.

been. **Don't ever do this without advice from a statistician!**

4.4 Inferential Statistics: Testing for Significance

Moving on to the reason why you performed your experiment; to make conclusions (inferences) from your data. Is the effect caused by your treatments real? Are your variables associated with one another? To answer questions like these, you need *inferential statistics*. This book would double in size if I were to discuss everything about stats. My goal is to help you to start. This section contains the key concepts that will make your life easier to know. After reading this, you should be better equipped to identify what you need to investigate further when it comes to processing your data (i.e. what to Google). It will introduce you to some of the phrases that you will come across on a statistics course or in a pure statistics textbook. Don't be too concerned if not all this chapter makes sense first time through. I've been doing stats on my experimental data for years, but I still frequently refer to wider literature to look up which tests to use and how to run them. The target is understanding enough to know what to look up.

Key Terms and Concepts

Confidence and Probability

The reason we perform inferential statistical analyses is to determine how *confident* we can be that our findings are real. The values that the tests generate tell us how likely it is that we would have obtained these findings by chance or, more specifically, the individual likelihood of obtaining a false positive or false negative. The chance that our experimental result isn't a real reflection of biology is the function of the variability in the system we are studying, imprecision in our experimental measurements, and the true magnitude of the effect being investigated. Increasing our sample size, changing the experimental design to reduce variability or improve measurement accuracy will increase our confidence that the results represent the real-life situation.

Your readers will want to know the confidence they can have in your findings, so we report these probability functions by using either a *P value* or power statement.

P values: testing for false positives (type I errors)

Often you will see data described as indicating a *statistically significant* difference or correlation. When you see statements like this, it means that there is a low probability that the observation being reported is a false positive. How low a probability does it have to be for it to be classed as significant? Well, that is should have been decided upon before the experiment when deciding on sample size. It was a subjective decision about how confident you would need to be happy to reject the null hypothesis.

The actual probability from the real data is the *P* value (sometimes written as p-value or *p* value; when writing up your work, check the instructions for the journal or assignment). Whereas data can only be significant or not significant relative to a line in the sand, the *P* value is a continuous scale, which is more informative about how confident one can be.

A *P* value of below 0.05 means there is less than 5% chance of this result being a false positive, while $P = 0.01$ means a 1% chance. A *P* value of 0.051 is very similar to

0.049, you aren't suddenly a lot more or less confident in your result due to being either side of this arbitrary line. Indeed, there is a strong movement for scientists to stop talking about the significance of their findings in this way and instead to just report the absolute *P* values on a continuous scale. The general advice, therefore, is to report your actual *P* values and describe your data through the framework of the effect size combined with your confidence. The one time this advice doesn't hold is when a *P* value needs to be below a certain threshold for another event to occur, e.g. "we need to be at least 99% confident that the drug works before we can prescribe it to patients".

A second very important point. Always remember that a *P* value that is above your critical threshold *does not mean that there is not a difference*. What it does mean is that you can't be *confident* that the observed difference has not occurred by chance.

Power: testing for false negatives (type II errors)

If you want to make a statement regarding how confident you are that there is "no difference" or "no correlation" between your groups, you don't use *P* value; you use *power*. This is the statistic you should report if you

> **Big Tip**
>
> P>0.05 does not there is no difference, it means that you can't be confident that the differences observed are due to true effects.

want to make a statement such as "treatment X *did not affect* Y". Testing for *non-inferiority* is slightly different from trying to determine if a difference is real or not. Definitely look it up or ask for help. Good news is that this relatively rare, but it might be needed if your hypothesis was relating to an absence of side-effects of a new drug.

Testing for normality

The first step in determining which stats test to use is usually to *test for normality*; this is the formal test we touched on earlier to determine if your data fits the classic bell curve distribution. If the result of that test is that your data are non-normally distributed it will affect your decisions about what to do next. What happens if your data fail the test? Well, don't worry! There are answers to these situations.

Transformations

Usually, the first step for non-normal data is to try a *transformation*. This involves making an equivalent change to all your data points in an attempt to bring them to normality. For example, you might take the logarithm of all your data points and then test for normality again. That first transformation might not work, so you could try something else. There are literally hundreds of transformation options, and it would take forever to test them all manually, but, the good news is that you don't have to... your stats program can do that for you. Look up how to do a *Box-Cox* in your program. This tool will test a range of different transformation strategies and help you to identify the one that will give you values that are closest to normal.

Once you have transformed the data, you use the transformed data in your statistical test. If it turns out that even after using transformations, your data is still not normal, it's not a disaster, but it means that you will have to use *non-parametric* tests.

Parametric and non-parametric tests

A working definition of these two terms is that you will use *parametric tests* to test normally or symmetrically distributed data sets (comparing means) and *non-parametric* tests when your data are non-normally distributed and you are unable to transform it to normality (comparing medians). All statistics tests make assumptions about your data to then infer the likelihood of the observed values arising by chance. When you can use a parametric test, it means that the data are more predictable and so the test will return a lower P value for an equivalent sample size. From a practical perspective, this means that it is worth spending the time trying to transform your data so that it is eligible to use the parametric test.

Usually the first step in data analysis is testing for normality

Degrees of freedom (d.f.)

The *degrees of freedom* in a statistical test is the number of independent pieces of information that went into the test. Unless you have some complicated experimental setup, the **d.f. is the sum of the number of samples per group minus one**.

For example, if you are comparing ten transgenic mice against ten non-transgenic control animals, then degrees of freedom are (10-1) + (10-1) = 18. It does get a bit more complicated as you move into bigger groups with more comparisons. However, your stats program will report this value for you as part of your results readout.

Multiple Comparisons and post hoc tests

This is a really important concept. Your inferential statistics test tells how likely it is you will have obtained a false positive or false negative. When you do one experiment, measuring one thing (one outcome variable), then the statistical tests are very simple, and the P value is easy to interpret. However, as you add more comparisons or more variables, then your statistics test needs to take these extra comparisons into account. The more things you compare, the greater the probability that you will get some false positives. We touched on this point from the other side in the experimental design part, where I recommended reducing the number of questions asked or selecting a primary outcome variable as a way to keep the sample size manageable.

The P value reported from a test tells you how likely it was that you had obtained a false-positive result. $P = 0.05$ means that there is a 1 in 20 chance that your result was a chance observation that doesn't reflect the real population. If, rather than comparing just one variable, you also compared four and then performed four *separate* stats tests, and those tests reported $P = 0.05$, this would mean that there are actually 4 x (1 in 20) chances that you have a false positive. If you kept going, comparing things in same experimental units until you finally had 20 tests, you would actually be more surprised if you didn't get at least one false positive! In these cases, when we are doing multiple comparisons, we need to perform a correction to allow for the additional chance of getting a false positive.

There are two ways that you can deal with multiple comparisons. First, and likely easiest, is to choose a stats test that builds in the multiple comparisons (e.g. an ANOVA accounts for multiple T-tests), and use a *post hoc* test to calculated the *P* value. A second option is to manually adjust your *P* values using an accepted correction technique (search for things like *Bonferroni correction*).

Post hoc comes from the Latin meaning "after this". When you are comparing multiple groups within the same test, the analysis happens in two stages. The first test run determines if *any* of the compared groups are statistically significantly different from one another known as *family-wise significance*, considering the multiple comparisons. You then perform a *post hoc* test to determine which pairs or groups are different *pair-wise significance*.

Whether you use a dedicated test or manually adjust, there are choices for which post hoc adjustment to use. Some corrections are more conservative than others and there are different requirements that must be met depending on how your data are distributed or how samples are grouped that define which post hoc option can be used. Therefore, whatever method you choose must be made clear in your methods sections of your paper.

The third option is not correcting for multiple comparisons. If you do this, you **will** overestimate your confidence in your interpretation of any individual result. Therefore, you must be very careful how you word your results and conclusions. If you do decide to go down this third route, always acknowledge that you are reporting the *unadjusted P* value in your results so that the reader knows exactly what you have done. In big data experiments, e.g. of the -omics type, it is quite common to use unadjusted *P* values and then report them as a list. The lower the *P* value, the more confident that you can be that the effect is real. This list can be used to generate a *false discovery rate.*

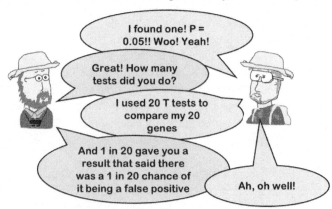

Remember to account for multiple comparisons in your stats test choice and when reporting your data

P hacking

The problem of failing to account appropriately for multiple comparison testing has coined a new term *P hacking*. This is the process of re-analyzing the same data on more and more features until you finally find one that reaches your critical significance threshold. For example, if you designed a clinical trial to test your new wonder drug but after analyzing all the patients as initially planned it turns out that the difference between the treatment group and non-treated group was not significant, you would likely be disappointed. However, if you then said, "Let's look at the effect on only the women in the study and see if there is a difference there", then proceeded to do a stats test where you only considered the female participants (ignoring that you even collected the male data), the *P* value returned might be lower but would be artificially so. If it still wasn't low enough low and you then went on to analyses whether just the people with blonde hair and blue eyes who were over 175 cm tall showed a positive response. Then people would make fun of you for being ridiculous.

This example takes it a step too far, but *P* hacking is now acknowledged as a real problem in the scientific literature. The desperation to find *any* statistically significant association is so strong that people will analyze data to the *n*th degree. Editors and reviewers are aware of this issue, and hopefully this will mean it will happen less frequently.

Remember that the review process doesn't end at publication and more and more papers are being retracted post-publication as a result of *P* hacking (either accidental or deliberate).

Investigating correlation? Consider the residuals

The *residuals* will be relevant to you when you are doing a correlation or line of best-fit analysis. The *residuals* are the values that come from the differences between each of the measured values and the line of best fit applied to the data.

How the residuals are distributed indicates whether the equation you have used for your line of best fit is appropriate. You perform a test for normality on the *residuals* before deciding on which regression analysis to perform (more on this shortly).

Identifying the appropriate stats tests

The processes you go through to identify the appropriate test to perform on your acquired data is very similar to what you did to calculate your sample size. The difference is you are working on a slightly larger number of options as you now have the information about spread and distribution.

Step 1: Identify your variables

- Categorical or continuous?*
- How many outcome variables were measured?
- How do you want to compare your independent variables? Split by factors or treatments?
- Decide if you are comparing relative to a reference population or directly comparing populations?

Step 2: Test for normality in your data set.

- Transform if necessary.
- Test for normality in the residuals for lines of best fit.

Step 3: Pick an appropriate test

- Identify the appropriate base test using the table on the next page or the flow chart at the back of the book (section 10.3).
- Look up the requirements of that test and confirm that your data meet those requirements *OR* select a modification from the base test if required.
- Run the test using your preferred stats program.

See section 2.7 (page 75) for a reminder of terms

Common stats tests

Independent Variable		Outcome Variable			Test
#	Type	#	Type	Readings /sample	P = Parametric NP = Non-parametric
1	Cat	1	Cat	1	Chi-Square Small sample size = Fisher's
1	Cat (2 groups)	1	Con	1	P – Independent sample T-test NP – Mann Whitney U test
1	Cat (2 groups)	1	Con	2	P – Paired T-test NP – Wilcoxon Matched-Paired
1	Cat (3+ groups)	1	Con	1	P – 1-way ANOVA NP – Kruskal-Wallis
1	Cat (3+ groups)	1	Con	2+	P – 1-way repeated measures ANOVA NP - Friedman's ANOVA
1	Con	1	Con	1	P – Pearson's Correlation / Regression NP – Spearman's Correlation or Kendall's Tau
2+	Con	1	Con		P - Multiple Regression
2+	Cat	1	Con	2+	P – Factorial Repeated measures ANOVA
2+	Cat	1	Con	1 and 2+	P – Factorial Mixed ANOVA
2+	Con/Cat	1	Con		P – Multiple Regression ANCOVA
1	Con	1	Binary	1	Logistic Regression
2+	Con/Cat	1	Binary	1	Logistic Regression
2+	Cat	1	Cat	1	Loglinear Analysis
1	Cat	2+	Con		P- MANOVA
2+	Cat	2+	Con		P - Factorial MANOVA
2+	Con/Cat	2+	Con		P - MANCOVA

Cat = Categorical, Con = Continuous. Parametric = normally distributed

Notes and other tests

- T-tests can be "1 way" or "2 way", use 1-way when the data can only go one way (i.e. only possible to go up or only possible to go down), use 2-way when the data could have moved in either direction.
- If you are comparing against an internal reference point (e.g. "normalised" data), then use the "1 sample" version of the test. This will consider the fact that, after normalisation, you will have no variability in your reference group.
- For time to event (e.g. survival analyses), usually, use a *log-rank* test

Post hoc decisions

Once you have performed your main test, you next do any necessary post hoc tests: For ANOVAs:

Bonferroni most conservative estimate of P values.

Dunnett's compares every value against a control but not against each other. Otherwise:

Variances	Ns per group	Recommended
Unequal	Any	Games-Howell
Equal	Equal	REGWQ or Tukey's HSD
Equal	Very different	Hochberg's GT2
Equal	Slightly different	Gabriel

Important Note

There are many other statistical tests available. Also, there are further variations of these common statistical tests to account for things like small sample sizes.

If your data do not fall neatly into any of the categories described above, then you may need a different test. I recommend using the table to identify what base test you should use and then you should look up that test and carefully check that your data fit the criteria associated with it or whether one of the modifications would be more appropriate.

4.5 Writing about Statistics

You've fought the stats battle; now you must tell people about your findings. In my experience, new writers often struggle at this point. I have highlighted the important stuff here for easy reference. Deeper and contextualised instructions are also included in the relevant writing sections in later chapters.

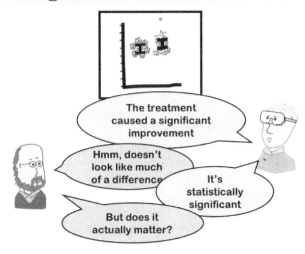

Methods section

Usually, the best approach is to have a dedicated sub-section within

Don't *only* think about the statistics, also consider the real-world relevance of your findings!

the materials and methods where you deal with all the aspects of data analysis and statistics. Here you should include all the details a reader would require to replicate and to understand your approaches.

Important things to include:

- How you defined the experimental unit.
- How many repeats were performed (biological and technical).
- How you tested for normality.
- Which transformations were applied (if necessary).
- How you dealt with outliers (if necessary).
- What inferential statistical tests were performed.
- How you dealt with multiple comparisons / post hoc tests.

Figure legends

State what descriptive statistics you have indicated on the graph (e.g. lines indicate mean ± SD). Also indicate in the legend the stats test that were performed on any graphed data, the sample size (N) and critical thresholds used for assigning statistical significance (e.g. * denotes significant difference from control, $P < 0.05$).

You do not need to include any of the actual descriptive statistics values in the legend; those numbers go in the results.

Results

Two rules here:

- Data in a table, no need to repeat the numbers in the text.
- Data in a graph, you need to include the relevant descriptive and inferential statistics in the results text.

Reading sentences filled with numbers is hard work for your reader and is likely to disrupt the flow of the information. Therefore, the best way to approach delivering the description of the important values is to focus the sentence of the overarching *story*, the major finding, and then put the numbers in parenthesis either at the end of the sentence or at the appropriate point in the sentence.

Wow, that's not fair!

Too scary!

P.06

Remember that if you want your readers to compare the values from one population to another, it is easier for them to do so if

Don't be afraid! It is not a disaster when your data don't cross the statistical significance threshold. The difference between 0.06 and 0.05 is very small in terms of how confident you can be about your findings.

those numbers are side-by-side. Therefore, my preferred option is to have a single set of brackets that contains the controls or comparison groups next to the treatments.

Example: Reporting stats

How you assemble the sentence changes the emphasis:

Option 1 – Emphasize the effect – preferred

"There was a two-fold increase in expression in population A compared with population B (mean A 20 SD 3, mean B 10 SD 2, $P = 0.02$ two-tailed T-test 14 d.f.)."

Option 2 – Emphasize the confidence – less common

There was a statistically significant difference in expression levels between population A and population B (mean A 20 SD 3, mean B 10 SD 2, $P = 0.02$ two-tailed T-test 14 d.f.).

Both options tell the reader the same information; they both contain the outputs from the descriptive and inferential statistics. Note that in option 1, the word "significant" does not appear, but the P value is reported, the readers can directly interpret the confidence level.

It is very rare that the confidence in the finding is more interesting than the biological effect. When your values do not reach your chosen threshold for statistical significance, you can change the phrasing to reflect your confidence. A third option is to write two sentences; one for descriptive stats, one for inferential, but that makes your writing which isn't a good idea.

Note: include the type of test performed and relevant statistics (degrees of freedom) in these sentences.

Chapter 5: Graphs, Tables and Figures

5.1 Figure Preparation
General information on how to prepare data figures for publication.
Including rules for sizing and standardization.
5.2 Tables
When to use a table and when a graph, and how to format tables so
that they are ready to publish.
5.3 Graphs
Help in selecting the most appropriate type of graph for your data,
and how to prepare your graphs for publication.
5.4 Figure legends
What should (and should not) go in your figure legends or captions.

5.1 Figure Preparation

Figures quality is really important!

Figures are the graphs, images and other forms of data presentation that goes along with the text to make it easier for your readers interpret the findings. In the early days of science figures weren't necessary for many things as there was a general trust in the report and because it was harder to print things like images effectively. However, nowadays your

> **Big Tip**
>
> Take pride in your figures. They will be one of the first things your readers and reviewers look at and will influence whether your readers believe your conclusions.

figures are effectively the evidence to support your conclusions. When it comes to your project report, manuscript, dissertation or any other form of writing that relies on primary data presentation, your work *will be judged on the quality of your figures*. The figure quality says something about you and your work: poor figures = poor quality data, messy = sloppy work, hard to interpret = something to hide. Spending time to make your final figures as good as they possibly can be is an absolute must if you want to be respected as a scientist.

The good news is that making a professional-looking, clear and accurate figure is a rewarding process. Making the figure can help you to see your story come together, and gives you satisfaction from feeling closer to publication or completion of your degree. You get a small win, a sense of accomplishment that will encourage you to keep on going.

Think about the figure before you perform your experiment

You know what your experiment was designed to determine, therefore you should also know what evidence you need to test your hypothesis.

With the experiment planned, you know the type of data you will obtain; numerical data, images, blots etc. Use this information to plan the figure before you begin. Think about what evidence you will need to include and how best to present it to deliver the message. Also, think about alternative interpretations or outcomes and how to rule out these alternatives. Thinking about what else *could* be going on before you start experimenting will help you make sure you capture all the data you need and not forget something.

Thinking about the figure before starting is most important in experiments where you need the data

Finished the experiment? Make the figure!

ordered in a specific way. The classic example is a western blot, where the load order on the samples will determine how you present the data. You want the samples loaded in a way that makes the interpretation easiest for the reader. Adding a gap between your control samples and test sample in your western blot might make sense for running purposes but do you actually want that gap in the final figure?

Make figure panels as soon as possible after analysing the data

As soon as you've processed the data and analysed the stats you should make the figure. I can't emphasise this enough. The experiment is not finished when the data is acquired, it is complete only once the figure is made (and the results written). On a personal note, when I was a postdoc I had a policy of not going home until I had assembled at least a draft figure. Importantly, the process of making the figure will help to identify if there are any problems, missing data or controls, or further experiments that are required. If you make the figure close to the time you are conducting the experiment then you will have all the cells, reagents, mice lines, etc. available if needed to repeat, extend or modify parts. Having figures prepared early will also make your lab meetings, talks and poster presentations quicker to prepare.

Data Presentation Rules

> ## Don't cut anything out!
> ## Don't add anything!

There are rules that you must abide to in your data presentation. The gross misconduct problems of making things up or deliberately mislabelling your data are clearly illegal and would have severe consequences for your career. However, and this is important, *any* massaging of the data to make it fit your story is also wrong. This includes deliberately omitting any part of the data or cropping parts of an image to remove parts you don't like to make it look more pretty.

Biological systems have variability, and the reader needs to be aware of that variability to be able to interpret your work. You must show the noise and non-specific signal. This means no photoshopping to make the perfect image. See box on the next page for the rules.

Two sub-tips:

- **Always report everything you have done**. Within the figure legend or methods section you should provide the details of any manipulations that you have performed. Give everyone reading your work the chance to evaluate your data fairly. It is hard to go seriously wrong if you are open and transparent about everything.

- **Seek advice**. In case of figures, all the reputable journals have a document that tells you precisely what you can and cannot do, read this carefully if you are ever in any doubt. However, same as everything else in your career, if you are unsure about something, ask.

Figure preparation rules*

- Do not add to, alter, enhance, obscure, move or remove a specific feature of an image: the focus should be on the data rather than its presentation (e.g. do not 'clean up' backgrounds or remove/obscure imperfections and non-specific bands).

- Adjustments should be applied to the whole image so no specific feature of the original data, including background, is obscured, eliminated or misrepresented as a consequence. Any alterations, such as non-linear adjustments (e.g. changes to gamma settings), must be disclosed in the appropriate figure legends and in the Materials and Methods section.

- The splicing of multiple images to suggest they come from a single micrograph or gel is not allowed. Any grouping or consolidation of data must be made apparent with dividing lines or white spaces and should be explicitly indicated in the figure legends.

- The same data in whole or part should not be presented in multiple figures unless explicitly stated and justified

- Previously published data in whole or in part should not be presented

*adapted from Journal of Cell Science. Read the whole thing here: https://jcs.biologists.org/sites/default/files/Revisionattachment_JCS.pdf

General Figure Preparation Comments

Different journals or publishers have different requirements for how figures should look, but many of those requirements are actually very similar. Differences are usually in the letter used to label panels and the dimensions of the figure. If you know where you plan to submit, you can look these up in the *instructions for authors* part of their website.

Figure preparation software

There are lots of software options which can be used to assemble your figures. The main requirement is a program that can produce high-resolution outputs in whatever format the journal requires (usually .tif, .pdf, or .eps). There is no point in taking a series of beautiful, high-resolution photomicrographs if they end up being too pixelated to make out the picture when sent to your reviewer.

The primary data outputs you start with will likely come from the program you used to acquire the data. For example, graphs will come from SPSS, GraphPad, or R, whereas microscope images will come from the software driving the microscope or after post-acquisition modification in programs such as image J or FIJI. Lots of your other data will be digital too and will come from different source programs.

To assemble the figure, you will need to bring these different formats together into a page-layout software. Within that software you may to need to relabel certain aspects to meet the journal requirements and to ensure consistency between panels. For journal submissions, your figures are usually uploaded separately from your manuscript files. This means you do not need to embed them in your word processing document. Indeed, it is much easier and will give you better outputs if you use a program designed for illustrations.

In terms of page layout software, Microsoft PowerPoint and Apple Keynote are quick and easy to use, but there **can be issues with the resolution of the outputs** so are generally not recommended. If you do use them, make sure the final outputs will meet your journal requirements. You will be able to generate better quality outputs using dedicated page-layout software. Ask your lab/supervisor what they recommend. My preferred programs are Adobe Illustrator or Corel Draw; however, both are quite expensive but your lab or Uni might provide a licence. There is also a free program called Inkscape that provides much of the same functionalities. As you would expect, there are learning curves with each of these programs, but if you are planning a career in science now is probably the best time to learn one of these better options. Producing good figures will continue to help you throughout your career.

Consider individual experiments as panels within a larger figure

If you are writing up a short research project, you might have data from only a few experiments to report, and it will likely be OK to have one figure per experiment. However, in larger-scale projects, your story will be built up from many more experiments that need to come together to complete the picture (remember the point about triangulation of experiments). When it comes time to publish, most journals have a limit to the number of figures you are allowed per manuscript and so each complete figure will usually be comprised of many smaller parts, we refer to these as *panels*.

Big Tip

Focus on getting each individual data panel perfect before assembling the entire figure.

When you are getting ready to submit your manuscript, you will assemble your panels into figures. Some panels will make it into the main figures, and others might end up in supplemental figures that will only be available online, while more again might only appear in longer format writing like theses or dissertations. Irrespective of where the panels end up, the quality should be of the same high standard.

At first, focus on getting each individual panel right, making sure that they individually show what they need to, that they are labelled fully and look as good as they possibly can. Working this way, you are less likely to make a sloppy mistake. You will also be able to finish the panel earlier than the whole figure, which means that you can complete an experimental series and get the satisfaction of ticking it off the to-do list. This will keep your boss happy, especially if they decide to use parts of your work as data to support a grant application or in one of their talks.

When it comes to putting together the final figure, thinking about the data as panels rather than figures means that you can easily pick-up and move things around and will be easier (psychologically) than if you have spent a lot of time carefully assembling a figure that you then have to re-assemble to insert one piece of missing data. This sort of late adjustment happens quite frequently.

When you are making the panels, you should standardise fonts, colour schemes and sizes of elements early (more on the specifics of this below).

Explore different ways to present the data

The clearer your figures are, the easier it will be for your readers to interpret the data, and the more rapidly they will assimilate. Before beginning, have a look at some published papers that have presented similar types of data. You should be reading these papers anyway, but pay extra attention to how each figure panel is assembled and identify what you think works best in terms of delivering the information. There isn't only one correct answer and you might be able to do better than the published versions, but the published examples give you a good place to start.

Your goal is to find the best way possible to deliver the points that you want to convey. Usually, this means making choices between different graph types and what data to graph vs present as a table, the different magnification level of images, whether to include a montage image or zoomed in box etc. What you should do in the end is dependent on what works best for your message. When you are crunching your numbers and working with your data, try out some different options and keep them as ready to go figure panels. Be aware that exploring data display options might also reveal aspects of the data that you otherwise would not have noticed. I also strongly advise getting input as frequently as possible as you work.

Your figures should be self-explanatory.

Labelling

Continuing the theme of trying to make your figures as easy to interpret as possible...make it your goal that the whole figure can be broadly understood *without* your reader having to read the legend. Looking at the figure, then down to the figure legend, then looking back to see the data, then down again to work out the next part then back up again is frustrating. Even reading that sentence is frustrating. The last thing you want to do is annoy your readers.

You have the ability to choose for the reader to not have to hunt for an explanation. All you need to do is use effective labelling. I'm not talking a crazy amount of added text here, just enough to ensure that the figure can be understood entirely on its own. If you label well the figure legend need only provide some extra detail that is relevant to the interpretation.

An obvious example is instead of labelling columns on a graph as 1, 2, 3 you can use short, interpretable version of the sample name. You can also write the antibodies, names of cells or treatments directly onto your images. Compare the left and right version of the same figure on the next page:

Unlabelled figure panel: reader has to find details in the figure legend to understand the data

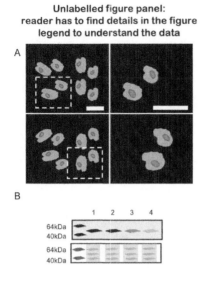

Adding treatments, cell types antibodies names etc., directly to the figure will help the reader see the message

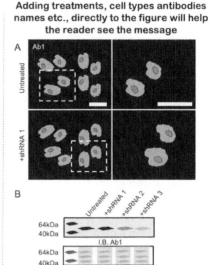

Whenever you can make things easier for your reader, marker, or reviewer, you should!

Representative images

Using representative images is a common way of showing an example of the message you are trying to convey. Be aware that whenever a reader sees something labelled as "representative", they will start thinking about how the writer decided what was representative of the population. Is it the best example? The most dramatic version of the phenotype? Or is does it actually represent the most frequently observed phenotype? The answer should be the last option.

When you have quantification to go with your image, use the central tendency of your scoring to identify the image you should use. If you don't have quantification, choose an image that truly represents the population and detail how they were selected, i.e. what they are representing, in your methods or figure legend.

> I'm going to use this picture of Idris Elba to represent me

> Perhaps a bit of a stretch?

Make sure you select representative images that accurately reflect the central tendency of your study population

Standardisation and Sizing

Professionalism in your figure preparation shows how seriously you take your work. An easy first step to help you achieve that "pro" feel is to standardise as many things as possible in your figure preparation. Keep the simple things the same from one figure panel to the next, including font choice and sizes, colour schemes, line widths, sizing of standard elements such as Y and X-axis lengths. This will help to make the whole body of work feel like a single entity that has been prepared with precision and care.

My lab has a set of guidelines that sets the standards for our figures, and then everyone in the team follows these guidelines when they make their figures. By having these standards, if one person makes the panel for Fig 1A and another the panel for 1B, we can merge those panels quickly and seamlessly. Ask your lab colleagues if you have guidelines too. If not, consider making a set.

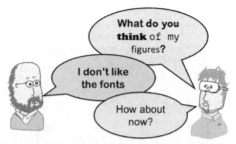

Keep consistent with your font choice throughout all figs. Arial, Times or symbol and don't use more than two sizes per figure.

Irrespective of whether you have a formalised lab guidelines document, ask for the figure files from the most recent manuscript your lab published (not the final tiffs/pdfs but the original files from the page layout program used to generate them). Use these files as a template to identify the standard fonts, colours, sizes etc.

Standardization Recommendations

- **Fonts:** Arial for standard text. Symbol where required. Courier if you need an equally spaced font for things like sequence data.

- **Font sizes:** Only use two sizes; 12 pt. for labels (A, B, C) and 10 pt. for everything else.

- **Lines:** 1 or 0.75 pt. thickness for every line. This includes all axis on graphs and any lines around or between images.

- **Graphs:** For most standard graphs (scatter, box and whiskers, etc.), a good size is height of 32.5 mm and width of 8.5 mm per column. This will feel small but trust me; graphs don't have to be huge.

- **Gels/blot:** My lab uses 6 mm width per lane; height varies depending on how they are cropped; again, set your own standards and stick to them.

- **File types:** Stay with the raw program file until the end, but make sure your program can ultimately save as .tiff if you use photos/micrographs or, if you intend to use vector graphics (line art, all graphs etc., .eps or .pdf.

- **Resolution and compression**: Try to keep everything at its native resolution as far as possible through the figure preparation process. Only compress if it is absolutely necessary. If you do need to compress, make sure to use a non-lossy compression system (e.g. LZW for tifs). This allows you the flexibility to scale up again if needed. Ultimately, you will save at 600 dpi or 300 dpi depending on if you have photos only or photos and vector graphics.

Colour Schemes

Five key points to consider here:

- **Consistency.** Use the same colour/symbol combination for the same treatment/cell type throughout your entire manuscript, thesis, or dissertation. This makes it easier for anyone reading the work to follow the treatment.

- **Cost.** Some print journals charge extra for colour figures. Consider using grayscale where colour does not add value. Also, even if you do use colour, print your figure in black and white to check things still work. Quite a lot of readers prefer the printed version of a paper, and they might print in black and white.

- **Don't use too many colours.** Two or three is usually sufficient. Choose strong colours with good contrast between them for different groups, different shades for within groups.

- **Colour blindness.** About 10% of the male population has some form of colour blindness. The most common is a difficulty in differentiating between red and green. Don't choose a colour scheme that is hard for 1/20th of your readers to interpret. The most common time when people forget this point is preparing images from fluorescence microscopy. The cameras or detectors on most fluorescence microscopes are black and white. This means that although you might have used red and green fluorophores to generate the images, the red or green colour that your 'scope is displaying is actually added after the initial acquisition as a *pseudocolour*. This means you have the choice or how to display your image.

~10% of males are red/green colour blind Don't choose to exclude some of your readers, reviewers or examiners!

The best option is to show single-channel images in black and white (maximum contrast), and show any merged panels in green and magenta. Converting red to magenta isn't a bad idea anyway; many monitors, projection screens and printers struggle with red on black, so using a brighter, more "contrasty", selection will help your images to look as good as possible.

- **Colour conversion.** Digital images are acquired in RGB colour schemes, whereas print images are generated using CMYK inks. Once you submit your manuscript, the journal will convert RGB figures to CMYK to print physical copies*. If you have a choice, you should be the one to do this conversion rather than them. This way you can be sure that the version you submit will look the way that you want it to in print form (always print your figure before submission). **Pro-tip:** Adjust your monitor to run on the same colour palette that your fig will be (Adobe RGB or sRGB) so that the way you look at it on the screen is as close as possible to the way it will appear in the journal.

This not an issue for online-only journals, you can use RGB images throughout.

Assembling the Figure

By this point you will have made a lot of individual panels, the next decision is how to assemble these individual pieces of the jigsaw into full figures, and how the individual figures can combine to tell a whole story. In the later sections of the book, we will come back to the writing process, but at this point it is good to have at least a very rough draft of the title of your manuscript and the main message you are trying to deliver. You need this to be able to make decisions about which piece of data fits where. Don't worry, it can and will change as you progress through the process.

Throughout the writing process, we will keep coming back to the same point; we want to deliver our information in a way that is easy for the reader to interpret, which makes the conclusions stand up and allows the value of our work to be appreciated. We often describe this as *"telling the story of the data"*. This is not say that this is supposed to be purely for entertainment or that you are working toward a big surprise twist at the ending, but rather that we are looking to have a coherent narrative that connects the work together into a fully connected single entity.

Now is the time to try and decide how you will connect the narrative. How will transition from one experiment to the next within the text?

"First we asked X and those data led us to ask Y".

In your results sections, you must discuss all the panels of every figure, and you must do so in the order they are presented. Indeed, the figure number and panel order establish the framework for the rest

Most journals have limits to the number of figures you are allowed. Group together the data from experiments that answer the same question so that each figure delivers a complete point.

of your paper; it's usually one of the first things I do when I am writing.

The order your present your work doesn't have to be the order the experiments were performed, instead, try to work out panel order where you can build the story and easily *flow* from one experiment to the next.

It is important to acknowledge that the different pieces of data are not equally important to the conclusions of the whole paper. Some panels will move the story along, others will provide essential proofs, others again will perhaps provide equivalent data in a different context or will rule out alternative interpretations. The way that you group your panels together and their relative location within figures gives you an opportunity to reflect this difference in relative importance. For most new writers, the tendency is to split things up too much, giving small, peripheral experiments the same amount of real estate and emphasis as the more important central story elements. Although this isn't necessarily a major problem, it might dilute your message and can lessen the impact of the whole body of work. Rather than have 30 separate figures with equal weight, you should decide what each contributes to the whole story and choose layouts and groupings to maximise the important parts.

Big Tip

Have a jigsaw session.

Try out different ways you can tell your story.

If you have a complicated, multi-faceted story, the process of deciding the order of elements and deciding which parts are core and which are supplementary can be quite tricky; however, spending time getting it right will make all the rest of your writing easier. If you are finding it difficult, one way that can help is to have a "jigsaw" session. This is exactly as it sounds. Print out each of your individual figure panels and then physically assemble those in a variety of ways to test out different ways to tell the story. This might sound like more effort than it is worth, but, trust me, it can be very helpful to visualise the story rather than talking about it in an abstract manner. It can help to tell your story out loud as if you were giving it as a talk. This can be done effectively by speaking to yourself but even better is to get your co-authors to listen. Getting the co-authors involved at this stage can save time in the long run as if you can agree on a message now you'll require less dramatic changes during subsequent editing stages.

Lumping or Splitting

Aim for each figure to advance the story by one strong, fully supported point. Size of the different elements will be important here, but let's look at the thought process of what to group together.

An easy first step is to put all the data from the same experiment in the same figure, for example, put your representative images and the quantification data from those images together. The next obvious way is to consider whenever you have triangulation data. Again, generally you should aim to put all those data together so that the readers will examine that work holistically and be more inclined to consider the weight of the evidence rather than be concerned about the individual limitations of each experiment in isolation.

Your readers won't appreciate having to magnify every image. However you decide to assemble your figure, make sure the final print sizes will work for your data.

141

> **Big Tip**
>
>
>
> Aim for 1 complete message per figure. Avoiding spreading too thin or lumping too much together.

Trying to do too much in a single figure can be a problem too. If you cram everything in you risk your reader missing something important. If you find you are getting to the stage where you are at panel X, Y and Z, it is probably time to send some material to the supplemental figures.

Don't bury crucial data in a small subpanel in the middle of a mega figure. This advice comes from my own painful experiences when the reviewers of one my manuscripts commented on the lack of data from an experiment that actually we had done and had included! It is not the reviewer's fault if they miss a panel that is critical to the whole story; it's yours for not drawing appropriate attention to it. Assemble your figs so that the important info is presented in a way that draws attention to them. If that means splitting a figure that you would otherwise prefer to "lump" then go ahead and do it, storytelling is about putting emphasis in the correct places. Use supplemental figures to cut down the complexity of a figure if necessary.

For a journal article or project report, think about how you will describe the data in your results section. Using one figure per subsection is an effective way to keep on message and make it easier for your reader to follow.

Size elements based on the type of data

This point sounds obvious, but you would be surprised how often I mark project reports that contain a single figure with a full-page bar chart or, worse, a tiny figure with 50 postage-stamp-sized staining images. When you are assembling your panels, print them out at different sizes. For journal articles, space will be at a premium, so you may have to try to identify the smallest size that still works. Always make sure that whatever point you are making is visible at the print size you have chosen. When going large, make sure that the resolution is high enough that it can printed at the size you want without looking pixelated. Remember that you might have an option of using supplemental figures to deliver any data that isn't central to your main story.

If you are writing a longer form document such as a PhD thesis or Masters' dissertation, you probably won't have as tight space limitations. This means that you can, in general, go larger, but don't go so big that it's ridiculous. If you do decide to go big, you should still plan to group together parts of the experiment where it is beneficial to the reader to have the different aspects side-by-side. If I am examining a thesis, I don't want to have to flip backwards and forwards from one page to the next to be able to look at the data I want to compare.

As you resize and restructure your figure make sure that you are keep to the figure preparation rules. Whenever you resize a panel, check that the font sizes don't change. If your fonts end up too small, the journal editors will bounce your manuscript back to you to be fixed prior to sending for peer review. This will slow down the whole process.

Perfect your alignments and spacings

Once you have selected which panels will make up each figure, you need to arrange them on a page. All journals have rules as to the sizing of figures. Journals that print in column format are likely to require figures to be either 1, 1.5 or 2 column widths. Set your page layout program to those widths before you go any further.

Now you have made beautiful individual panels don't ruin the effect by putting together the figure in a haphazard way. Assemble the panels in a neat and visually pleasing way that doesn't leave too much unused space between the different elements and, at the same time, isn't too cramped. Consider the sizing of elements again and make sure everything ends up appropriately balanced, at the right size for the data type and for the emphasis you want to place upon it (see cartoon below).

All page layout programs have alignment and spacing buttons: use these! Align the *locants* (the A, B, Cs that identify the panels), align the left-hand side of the panels. Align the titles and axis of your graphs with one another. Group your panels together and use the spacing tools to ensure the space between panels is standardised. This small handful of extra button presses to get everything squared-off really helps the feel of professionalism in your figure.

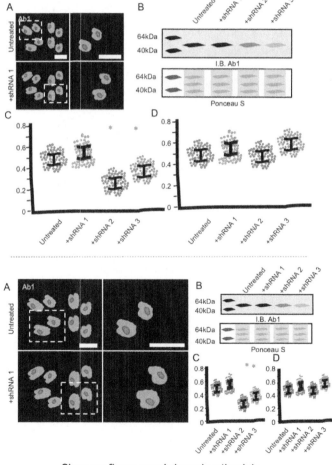

Size your figure panels based on the data.
Usually this means that images should be larger to show the
detail you are describing whereas graphs and blots can be
smaller to still deliver the same message

Print your final figure

- Print your figure; *make sure the sizes, colours and arrangements work.*
- Print your figure; *it's easier to see typos or alignment issues.*
- Print your figure; *it feels good to have finished the experiment*
- Print **all** your figures from the manuscript; *assessing them side by side will make it easier to spot inconsistencies from figure to figure.*

If you use a hard copy lab book, stick the printed figure/panel in your lab book with the details of where the electronic files are located and their file names, as well as where the original and processed data are located.

Get input from someone who doesn't know your work

Finished? Now's the time to get feedback. Give your figure(s) to someone who doesn't know your work. Although you will (ultimately) want the input from people at the professor level, you also want your work to be understandable to readers at every level; therefore, you can give your figs to a student in a different lab and you should be prepared to do this for others.

When you give someone a figure, you **shouldn't need to explain it to them**. If you have labelled it effectively, they will be able to understand without you saying anything. To test this, ask them to

Big Tip

Get someone who doesn't know your work to explain your figure to you. If they can't explain it, then you need to do more work.

explain the figure to you. This will give an impression of what a naïve reader will think when they see your work. If you are not happy with the way they explain it, then it's a clear signal that you haven't done as good a job as you thought you had. If they can't explain it at all, it's often a sign that you haven't labelled things effectively.

If you are happy with how your reader has interpreted your figure, next ask them to check for inconsistencies, alignments, labelling etc. All the small stuff really makes a difference in your readers' impression of your professionalism. In my lab, we discuss newly made figures together as a group during lab meetings. Having many eyes looking at the same thing increases the chance of spotting a mistake.

What to check when editing

Having read the info above, it will not come as a surprise to you that the most common problems are due to either inadequate labelling, sloppy alignment, and inconsistent sizing. During the editing stages, have another check over everything. Don't forget to think about *emphasis* and be prepared to change the panel locations if you think it will help to direct more attention to the most important elements.

Compare your final figure to examples from published papers in your target journal. Is your work of a similar or higher standard? Once you are happy with your figures, you'll be ready to write the rest of the manuscript.

5.2 Tables

Almost all research papers use tables and/or graphs to deliver numerical data. The next two sections deal with these two most common elements.

Table or Graph?

You shouldn't deliver the same information twice. Therefore, for each set of data you need to decide whether a graph or table is the most *effective* for supporting the narrative of the manuscript. This choice is one of the many ways you can choose to increase or decrease the relative emphasis on specific findings within your story.

As general advice, graphs are easier and quicker for a reader to absorb and therefore best for making an impact. However, graphs take up more space and don't deliver fine detail. In contrast, tables require the reader to do more of the work to find the parts that matter to them, but they are effective at delivering a lot of information efficiently in terms of space.

Big Tip
Don't use a table if you could write the same information in five lines of text or fewer.

Whenever you choose to use a table, you are making a trade-off; showing all the details means that the reader has access to everything, but you have to acknowledge that it is much harder to draw attention to any one thing, only the really committed readers will go line by line through the data. This means tables are best when you need to deliver large or complex data sets but when the *individual* points are **not** central to the story.

Tables are often the best choice for reporting the demographics of your participants or patients, for lists of clinical details, and for things like primer sequences for PCR. Tables are also commonly used for long lists of proteins, genes, miRNAs etc. coming from -omics or other big data sets, as, in these cases, each reader will have a different favourite gene or protein, and you can cater to all your readers' needs by presenting everything. Often these sorts of massive tables are best as supplementary material.

Tables – Points to note:

1. . If you can write out the contents of a table in five lines of text or fewer, then you don't need the table. Small tables are frowned upon by journals; they take up more space than necessary
2. Don't use tables in talks. Tables disrupt your flow as the audience will stop listening to you and try and read the table instead.

Writing about Tables

In your final manuscript text, you *do not* need to repeat any of the details that are delivered in a table. All you need to do is direct the reader to where they can find the information. If you think it will benefit your narrative, it is generally OK to highlight a small number of the specific entries that are important to the narrative, e.g. those that you intend to bring up in discussion, but don't do this too frequently.

Writing about tables examples:

1. Intron-spanning primers were designed to amplify LAMA3, RPLP0 and GAPDH (Table 1).
2. We compared the clinical characteristics of patients with and without autoantibodies (Tables 1 and 2). The frequencies of presentation with other allergic diseases were higher in patients with autoantibodies ($p = 0.002$), while malignancies were lower in the autoantibodies group ($p = 0.01$). There were no other significant differences between populations.

In example 1, there is no need for any more detail. Example 2 highlights the points of interest from a much bigger list; adding those extra phrase helps the reader identify and appreciate the importance those specific points.

Formatting Tables

Have a quick look at some published papers and you will notice that although there are subtle differences between different journals, their formatting is essentially quite similar and quite minimalist. Technically, you don't have to format your tables at all, the staff at the journal will do that for you. However, submitting work that looks like the finished article will add to the overall feel of professionalism as far as your reviewers are concerned. Similarly, in your thesis, project report or dissertation, you want everything to look to be of publication standard. Therefore, it is a good idea to adopt a style used by one the of bigger journals in your field. The generic formatting below takes very little time to achieve. No special software is required, you can produce tables in Word or Excel, LaTeX, Apple Pages or Numbers etc.

Table 1: Table title (descriptive + short)

Independent variable / grouping	Measurement Title 1	Measurement Title 2	Measurement Title 3
Row Entry 1	Data	Data	Data
Row Entry 2[1]	Data	Data	
Etc	Data	Data[*]	
Etc			

[1]footnotes, e.g. description of measurement, outlier etc. *footnotes, e.g. indication of significance level and comparison and stats test, e.g. $p<0.05$ relative to control, paired T-test.

Table formatting rules

- Top and bottom borders only, do not put borders on the left and right.
- One line between titles and main body. No lines between other rows/cells.
- Arial/Times, font size 10, 11 or 12 (may be defined by journal).
- Table titles are placed above the table. Use a short title, usually descriptive rather than declarative (less than one line).
- Generally, no legend required. The title and data are sufficiently clear on their own.
- Use footnotes to add specific details. Indicate footnotes with superscript numbers, letters or symbols, then define the symbol below the table.

Things to watch for in tables

- Use an appropriate level of significant figures. Do not use more significant figures than your measurement system supports. Look at the dispersion of your data to determine how precisely to report the central tendency.
- Indicate in table headings or footnotes whether the descriptive statistics are mean, median, mode, and whether the variability is SD, SEM, CI.
- Indicate in footnotes the statistical tests performed
- Indicate the "N" of the study population.

5.3 Graphs

I love graphs. One of the first things I do with new data is test different ways of visualising those data to identify trends and to find the way that works best for delivering the message. I also hate bad graphs and get really frustrated when I see poorly prepared and presented work being published. These feeling are very common for anyone used to working with data! In this section of the book, I will focus on how to make the best choice for plotting your data.

Choosing the Best Graph for Your Data

Quite simply, the graph you should use is the one which makes it easiest for the reader to see not only the core result but also any information required to interpret the data. Usually this means showing as complete a picture as possible. Remember that

Big Tip

Choose a graph type that gives your reader the opportunity to understand and interpret the data for themselves.

there are science ethics requirements for openness and completeness in data reporting. Together this means that you should using something *better* than a bar chart*. There are hundreds of graphing options, so the next few pages will help you to decide. As usual, we'll go step by step.

What type of question does your data answer?

Think about your question and what you are trying to tell the reader. Most of the time, you can classify your question into one of these five groups:

1. **Does your question ask about the *variability* within a group of data?** What is the range of the data, the shape, or the centre of the distribution? (see *Variability graphs*).
2. **Does your data *compare* two or more groups?** Treatment vs control, population 1 vs populations 2 and 3 (see *Comparison graphs*).
3. **Does your data ask if two populations are related to each other?** Does X increase as Y decreases? (see *Correlation and relationship graphs*).
4. **Does your data ask how a total is split between subgroups or compare how different populations are each split into subgroups?** What proportion of your population is male/female? How do ratios of X differ between treatments? (see *Proportion graphs*).
5. **Are you plotting time to an event?** Did your treatment increase survival times? Is the expression level of a protein related to patient outcomes? (see *Time-to-event plots*).

Quick reference guide on next page and at the back of the book

*Throughout this chapter we are talking about presentation in manuscripts or thesis. In posters, talks or public engagement-type activities the data may not need to be as tightly scrutinized and it might be more rapidly absorbed by your audience if you use a simpler graph.

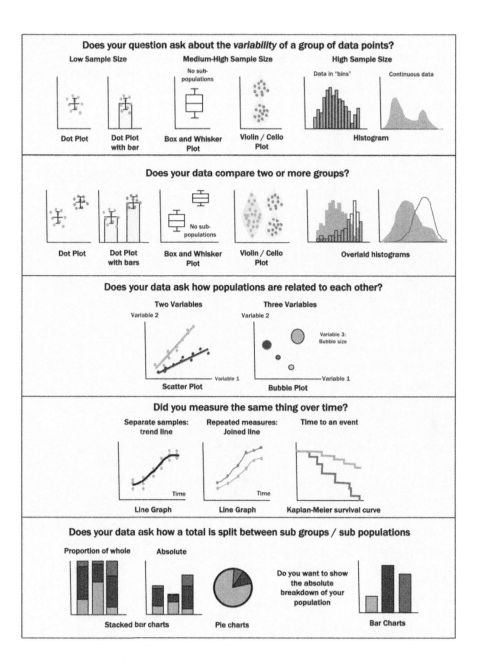

Averages and Error Bars

Very quick reminder. On most graphs, you will want to indicate your population summary statistics; but which ones?

For all of your subpopulation populations, look at the spread of the data around the mid-point and ask, is it roughly symmetrical? Are the mean and median values close together? If the answers are yes for *all* the populations you are plotting then indicating the *mean* and *standard deviation* are usually the best choice*.

If any of your populations are not symmetrical, look more closely at the distribution. If there are skew or large outliers, then a *median* with either a *range* (5th to 95th percentile is common) or *confidence interval* is likely a better choice.

If you have two or more subpopulations, then you should ask yourself does the central tendency of population say anything valuable about the population. Often the answer to the question is no and, in which case, there is no value in showing either the mean or the median: your storyline will not be about the average but rather the size or relative proportion in each sub-population.

Don't plot impossible error bars

Your error bars should never stretch into impossible ranges. In many experiments, it is biologically impossible to obtain values that are either below zero or above 100%. If you plot a graph with impossible errors on them, it immediately tells your reader that you haven't thought about what your data means and that you have used the wrong type of dispersion measurement!

If your standard deviations stretch into impossible territories it tells you that your data are likely to be skewed. It's probably time to look at medians and changing to 95% confidence intervals.

Rarely, it may be important to indicate the range within which the true mean lies (a confidence graph). If that is the story you are telling, usually the mean and confidence interval or range are the best options.

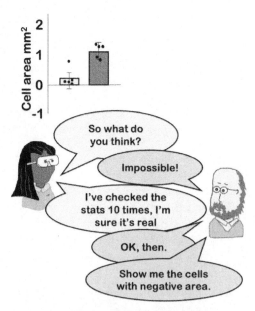

Watch out for impossible error bars!

Variability Graphs

Example question

"What was the range and distribution of X in the population."
Where *X* is a continuous measurement such as size, area, intensity, age, height, speed, blood pressure etc.

Most common options

- **Dot plots**: small to medium samples size (<~40 per group).
- **Box and whisker plots**: medium to high sample size (>~15 per group) but not good in situations where you have sub-populations.
- **Violin plots**: medium to high sample sizes (>~15 per group) and good for when you have sub-populations.
- **Histograms**: for large sample size (>~50 per group) and good for when you have subpopulations.

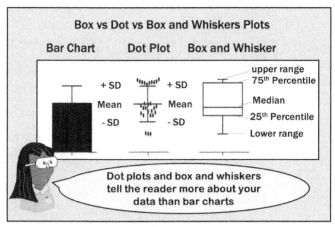

Dot plot (scatter plots): small to medium sample sizes

Dot plots are nice and simple, usually the first choice whenever you have categorical independent variables and continuous dependent variables. Every data point is plotted on the graph for all to see and, therefore, are great for a reader who wants to know about the distribution and variability within your samples. Any time when you have fewer than 10 points per population, you should always aim to show all the data, and even if you have up to ~40 points you probably should still consider a dot plot. However, you will find that the plot begins to become messier as the numbers go up and they can become more difficult to interpret. Once that happens the added value of a dot plot gets trumped by the loss of aesthetic appeal and its effect on reader experience. Try it out with your data and see what you think.

The dots on a dot plot tell the whole story. However, it usually helps your reader to also include a line at the mean or median, and to indicate the standard deviation or confidence intervals (in addition to showing the dots). If you like the look of bar charts, you can include in addition to the points. Leave the bar unfilled so that you retain the professionalism benefit of showing the data.

Box plot (box and whiskers: medium to high sample size

Most graphing programs have built-in features for turning a continuous data variable into a box and whisker graph. Here the traditional options are the box runs from the 25th to 75th percentile with a line at the median, and the whiskers stretch to either the range or to the 5th and 95th percentile. In the latter case outliers are denoted by * or ° depending upon how far they are outlying. The whisker options can be changed to whatever you prefer to display your data. As this is a choice, you must make what has been plotted clear in your figure legend.

Box and whisker plots are useful for showing the spread of the data but only really work when sample numbers are quite large. As a rule of thumb, box and whiskers work well need at least 15 points per population. However, and this is important, as you are only plotting the cut points within your population, box and whiskers only work when your data are spread reasonably evenly around the mid-point (doesn't have to be normally distributed). **If you have two or more distinct sub-populations then you should look for an alternative graphing mechanism;** a histogram if you have enough data, a dot plot if you have fewer points, or a violin plot if you want something a bit fancy!

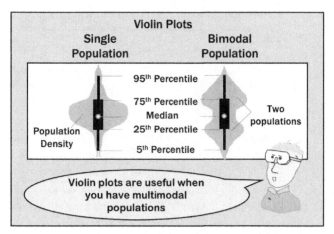

Violin / Cello plot: medium to high sample size

A slightly modified form of the box and whisker plot, the violin plot, is gaining in popularity. They might look a bit complex at first, but these plots are essentially just a mixture of a smoothed histogram turned on its side (termed a *density plot*) with a box and whiskers overlaid in the middle. A greater proportion of the study population are located in the areas where the "violin" is fatter. This graphing mechanism looks quite good all the time, it fits our key criteria of telling the whole story of the data. However, violin plots are especially effective when you have bi-modal or multi-modal populations. You can tell more of the story of the data to your readers with the single image, and as such, they can work well.

Most graphing programs will *not* have violin plots as a default option, but you can find free online scripts that will make them quickly for you if you decide that this graph is the best option for your work.

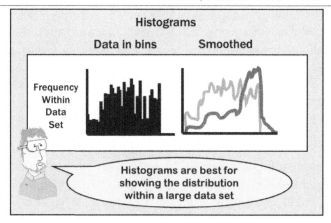

Histogram: large sample sizes

Histograms are used to plot the frequency of data occurrences within a continuous data set. In these plots, the data are split into *bins*, and the plot presents either the absolute abundance or the percentage of the population within each of those bins. When the sample size is very high, you have an additional option of plotting the histogram as a smoothed line. Quite often you will see histograms used for describing not only the dependent variables but also the independent variables, for example plotting the break-down of the ages of participants in your study using bins of 0-5, 6-10, 11-15 etc.

Histograms are good for when you have split or multimodal populations, e.g. a low group and a high group within the same population.

You will need a lot of data points for histograms to look good, the number required depends on the number of bins

Discarding data that you don't like is forbidden. You must report what you have acquired. However, using discrete bins can help make sense of a data set.

you use. As usual, there are no hard rules, but I would suggest a total sample size that is at least 10x higher than the number of bins that you plan to use. For example, if you wanted to plot the age ranges of your sample, and you had 80 participants, I would have a maximum of eight bins. When you are plotting more than one group (overlaid histograms) then choose the number of bins based on the population with the fewest samples, e.g. if you had 80 males and 65 females and wanted to illustrate the age distribution of both groups, then I would use a maximum of six bins. Obviously, fewer bins would be fine too! When deciding where to put your points, there might be a case logical positions rather than arbitrary; if you were thinking of ages then having bins of below 18. 18-65, and 65+ might make more sense if those groupings were relevant to the data interpretation. Of course, if you change the data in this sort of way, then you are also turning a continuous variable into an ordinal variable, and this might mean that different stats test are required. Bin cut points should be always be justifiable.

Comparison Graphs

Example question:

"Is population 1 different from populations 2 and 3?"
These sorts of questions are very common in science, and these types of graphs are likely to be the ones you use most frequently.

Most of the best options for comparison graphs are the same as for distribution graphs (no surprise), we are showing the same things except now you have two or more groups:

- **Dot plots** for small to medium sample sizes.
- **Box and whisker plots** for medium to high sample sizes.
- **Violin plots** for medium to high sample sizes.
- **Overlaid histograms** for high sample sizes.

With extreme caution…

- **Bar charts** but only when you *do not* have variability within your population (single measurements).

Should you plot absolute or relative values?

The comparative graph you are about to make could serve two purposes; first, it could deliver the data from each group in *absolute* terms or, second, it could compare *relative* values from the populations against one another. The decision about which option to plot isn't trivial, it is something that requires careful consideration. You might find that if you plot the absolute values then any differences between groups are hard to see; however, if you plot the relative values, you might end up emphasising a difference that is so small that it is biologically meaningless. So, what should you do?

We care about the true biology the most. Therefore, the recommended option is to show the absolute values whenever they are a direct biological measurement, i.e. something that connects with the real world such as age, height, enzymatic activity, drug cleared, migration speed etc. Use the relative values when the measurements don't equate to a direct biological phenomenon, when they rely entirely upon an internal measurement for context or calibration, e.g. quantification of indirect antibody-based techniques or fold change from ddCt quantitative PCR assays.

Whenever you decide to use relative values, make sure you discuss what those results mean in terms of the real biological context in your document. A 100-fold increase in something that was barely present might not make any biological difference.

Do not repeat things unnecessarily. However, sometimes, it can be appropriate to show both graphs (one graph may end up in the supplemental figures and one in the main text). Or you might justify recording the absolute values of lots of different outcome variables in a table and then include the relative values of a select few

Your goal should be to present your data in the best possible way, not just the bare minimum!

interesting outcomes as graph. This could be effective as way to highlight the differences between the populations that are central to your narrative. Be cautious with the double approach, only do it if definitely adds value for the reader. Of course, you will also describe the results in text form in your Results section. In the text, you can include both descriptions and add emphasis to one or other aspect using a graph. Remember that the figures will likely have the biggest influence on the reader, so make sure that the graph in the figure conveys the most important point.

Why not bar charts?

The *only* time when bar charts are the best option is when there is **no** variability in your data set. Examples are binary outputs or when a population is measured only once. In these situations, when the values are absolute, you are not hiding anything by using a bar chart. However, in most other situations you do have variability and there are almost always better options than the classic *dynamite plunger* chart.

Fundamentally, when you use a bar chart, you are making a choice to *minimise* the

Adding bars, or lines to your dot plots can help your reader see the trends in central tendency and dispersion. All the benefits of a bar chart without hiding the data spread.

information available to your reader. You are saying that they should accept everything rather than make their own decisions. As scientists, it's our job to question everything. Whenever someone isn't prepared to share the evidence, alarm bells start ringing. Although lots of bar charts are published, there is a definite move away from these incomplete graphing mechanisms. The burden of proof is upon you to show that they are the right type of graph. As the editor or a reviewer of a journal article, I will always ask for a better graph to be used.

Correlation and Relationship Graphs

> ## Example question:
> *"Does Y increase as X increases?" "Is there a relationship between variable 1 and variable 2?"*
> If your independent and dependent variables are continuous, you will need a graphing mechanism that will allow you to show both aspects to compare your study populations.

Three main options:

- **Scatter plots** for when you have *two* continuous variables and want to display how they relate to one another.
- **Bubble charts** for when you have *three* variables to compare at once.
- **Line graphs** when you measured the same thing multiple times and want to demonstrate how the measurements have changed over a course of time or after successive treatments.

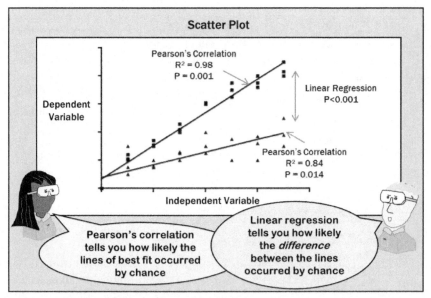

Scatter plots: two continuous variables

A scatter plot takes individual points in your data set and plots them on the two or three axes. A "line of best fit" is then added to indicate trends within the data, and you analyse these trends using *regression analysis* (more below). A scatter plot should be used when both variables are continuous; for example, plotting age, height, or weight against experimental measurements or biological responses.

Convention is that the independent variable is plotted on the X-axis and the Y-axis is used for the dependent variable. In other words, the outcomes that were measured are plotted on the Y-axis, while the intrinsic features of the population are on the X.

Lines of best fit – r and R^2

The reason you are plotting your data is to show the reader how the variables relate to one another. The line of best fit is your way of highlighting any trends that exists and the strength of any association. This can be a quite complex subject, but I have included a few core concepts below to help identify what needs to be considered or looked-up.

The classic, simplest, line of best fit is a straight line which passes through the middle of the data points in such a way that the number of points above the line and below the line, and the net distances to the line are approximately equal. You could calculate this location manually, but a more accurate and less laborious way of determining the location and slope of this line is to use the *least squares method*. This calculation does exactly what you'd expect: it identifies the position where the sum of the differences to the line is the lowest. Most statistical programs will do this for you, or you could use an online calculator. One decision you will make is if the line should be constrained at any point, the most common being when you know that at 0 on the X axis would be 0 on the Y axis in the real world. This isn't always the case, its something you need to decide for your question.

The least squares method generates the equation of the line ($y = mx + c$ where m is the slope and c is the intercept). However, it also generates a numerical indicator of how well the line models the data; this is the *correlation* value and is denoted r. Non-linear lines have more complication equations that describe the line, but also use r values to indicate *goodness of fit*.

r values range from -1 to 1. This number carries two important pieces of information about your line:

- The **strength** of the correlation: r = 1 or -1 is perfectly correlated, r = 0 no correlation.

- The **direction** of the correlation. Positive r numbers mean as one variable *increases* the other also *increases,* whereas negative r numbers means as one variable *increases* the other *decreases.*

A connected statistic, the R^2 value, indicates the *proportion of the variation that can be attributed to the independent variable.* R^2 is, as the name suggests, is simply the square of r and therefore is always between 0 and 1. As R^2 gets closer to 1 the model of the line is a better fit; the points are spread less far from the trend line (if $R^2 = 1$ then all the points will be perfectly on top of the trend line). This holds true whether a line is straight or derived from a polynomial.

Report the r or R^2 values in your results sections and directly on the graph.

Once you know that you can create a "goodness of fit" value for your line, it naturally follows to ask; what values for r or R^2 indicate that a relationship is *meaningful?* The answer to that is subjective (aren't all these things?); it depends on your study question and, to an extent, on the discipline. On the next page is a table of some of the norms in different disciplines. Don't take these as rules, as with everything relating to statistics, you must think about what *your* numbers mean with respect to *your* question and study population when you are interpret them.

When does an r or R^2 indicate a relationship? (rough guide only)

Discipline	r meaningful if	R^2 meaningful if
Physics	r > 0.95	R^2 > 0.9
Chemistry	r > 0.9	R^2 > 0.8
Biology	r > 0.7	R^2 > 0.5
Social Sciences[1]	r > 0.6	R^2 > 0.35

[1]Humans are harder to predict than chemical reactions; hence, lower strength correlations are considered to be meaningful in human behavioural studies.

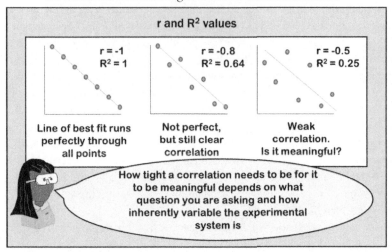

Significance testing and confidence intervals for slopes

The r and R^2 values of a scatter plot tell you how well the data fit a model, but they don't tell you about the *confidence* you can have that your model accurately describes the wider population. This is important. Just because your individual assay gave a high R^2, does not automatically mean that the association could not have occurred by chance. To report confidence, we will need a probability measurement, our old friend *P*. In this case, the *P* value tells how likely it was that the observed correlation has occurred due to sampling variation.

If you are dealing with a linear relationship, then the simplest test of whether there is a correlation between your X variable and your Y variable is *Pearson's correlation coefficient*. This test will calculate the r statistic, the *P* value, and also the confidence intervals. Everything you could ever want!

Check your residuals!

As with all stats tests, you need to check that your data fit the rules of the test. For case of correlations, you need to check the raw data and the *residuals*. In this context, residuals are the difference of each of your points *relative to the line of best fit*. By definition, the data points will be approximately equally spread above and below the line of best fit, the sum of your residuals will always be close to zero. However, their *location* relative to the X-axis might not be equally spread, and it's this we need to check (see diagram on next page).

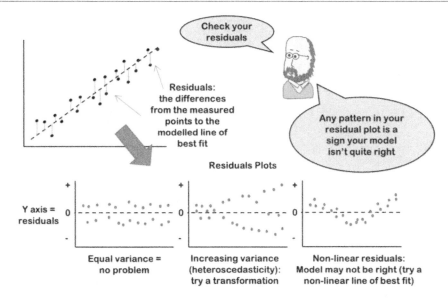

It's not just about stats, checking your residuals should be something that you routinely do whenever modelling a line of best fit.. Any time where there the residuals don't display equal variance, it is a sign that something is wrong.

If you have *heteroscedasticity* (increasing variance as you move in one direction) then likely something like a log transformation could help. However, if there is clear shape, it likely suggests that your line of best fit isn't right. Most likely, you will need to shift from a linear to a non-linear model.

$R^2=1$ and $P<0.01$, Can you say X causes Y?

Not necessarily...

Correlation does not imply causation,

Consider the contribution of confounding variables in your discussion.

Correlation does not indicate directionality

Consider the possibility of reverse causation in your discussion.

You must describe the results using the correct terms for what you did: if it was an correlative study you describe the data as "X correlates with Y."

Beware spurious correlations

Are my two slopes different from each other (*regression analyses*)?

Everything we have talked about so far in terms of correlation graphs is about accurately modelling the line of best fit of one population. However, if you have two or more slopes that you want to compare, then you need a separate statistical test to determine how confident you can be that any observed *difference* between the two (or more) slops is not a false positive. To compare lines, you use *regression analysis*. See the box on the next page as a starting point for finding what to look up.

If you are comparing slopes, you will end up with a separate *P* value for each of the models that describes each of your lines of best fit, *and* another, distinct *P* value for whether the lines are different from one another. The first tells you how likely the associations could have occurred due to sampling, the second how likely it was that the difference occurred due to a false positive. Report all of these *P* values in your results and/or put the r² values and *P* values on your graph.

Types of RegressionAnalyses

Looking at regression options in your stats program can be intimidating. Here's a list of some common terms to help you find the right one.

Linear Regression: Use to compare lines between two or more groups. Also known as *ordinary least squares* or *linear least squares*. Despite the name, this type of regression can also model curvature. Generally, this is the most commonly used, but linear regressions do have sensitivity to outliers and are prone to overfitting so if your data look a bit *off* it may not be appropriate. Another potential problem arise if you have multiple independent variables that correlate with each other; termed *multicollinearity*.

Ridge regression: A variant of linear regression that is designed to compensate for multicollinearity.

Lasso regression: A further variant of ridge regression that tries to simplify the model through selection of a variable to emphasise.

Partial least squares regression (PLS): Useful when you have very few observations or when your independent variables are highly correlated.

Non-linear regression: If you check your residual plots and the linear model has problems, then a non-linear regression will try a wider range of curves to attempt to find a fit. You might also see this described *polynomial regression*.

Poisson regression: Use when your dependent variable is counts, e.g. to assess the rate of occurrence. A variant of poisson regression is a *zero-inflated model*, where there might be more than one mechanism to account for there being a larger number of zeros in a data set.

Binary logistic regression: Use when the dependent variable is dichotomous (yes/no, alive/dead).

Ordinal / nominal logistic regression: Use when you dependent variable has either ordinal (low/medium/high) or nominal (giraffe/zebra/llama) responses.

Bubble charts: comparing three variables

On a bubble chart, you plot two variables on the X and Y axis as you would for a scatter graph, but then represent a third variable by changing the area of the bubble. This is usually used to categorise study populations into different groups by using coloured bubbles. Like the scatter graph, a bubble chart is used to compare groups and show correlations. They can be quite effective to show broad trends but gets messy when you have too many bubbles.

Bubble charts work best when your X and Y-axis variables can be split into bins, and when you want to show that there are larger/smaller proportions in the different bins between your comparison groups.

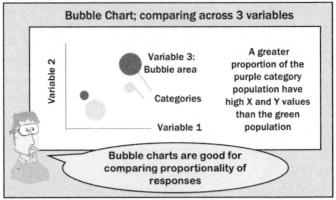

Line graphs: for connected variables

A line graph takes either the direct data or some form of summary statistic (mean/median) from the measured variable and plots how those values change through repeated measurements. Often these types of graph get used when you have multiple readings over a range of time, temperature, force or multiple different treatment types. Line graphs are good for showing trends within the data set and can be very effective in situations when you have many different groups/treatments and multiple measurements per group.

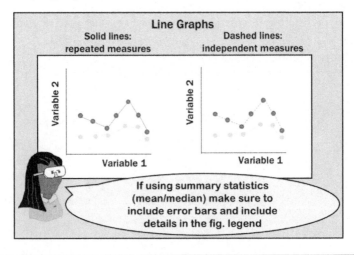

Solid lines or dashed connections? The convention is that solid lines indicate repeated measurements of the same experimental unit, e.g. if you took the blood pressure from the same person every hour or measured the deformation of a material as force increased, then you can connect the points on your line with a solid line. However, if each point comes from a discrete sample then you should use a dashed or dotted line to connect the points; e.g. if you had a dose response curve using six separate groups each with a different dose of your drug, or if you culled animals at different time points then a dotted line should be used to display the trend.

Quite often, experiments involve paired analysis; before and after an intervention. In these cases, while you could show just the before and after points in a dot plot or box and whiskers, showing the individual points linked by lines can be informative for about how consistent the results are. The per individual responses become important when you have spread data and detect only a small overall average increase in the population but when almost every member of the population showed the same direction of increase. Even if you don't intend to use a paired graph to display the data in the final manuscript, it can be useful for you to plot the data this way during the data analysis phase so that you can spot any individuals displaying unusual responses. The advantage of the linked-line type graph is that it carries the real-world information of the spread in the initial values, whereas the unconnected dot plot lacks this info. Of course, with a benefit comes a disadvantage; all the lines connecting the points can mean that the plot becomes messy and harder to interpret. As with the other graph options, my main advice is to try out different styles and then decide what is best based on what you want to reader to take away.

Before we move on, it would be remiss of me not to mention again that it is best if you can show the distribution of data at each point of your line. So, again, preferably show the actual spread of the real data but at bare minimum mean + standard deviation or median + 95% confidence intervals and remember to include in the details of what you have plotted in your figure legend.

Proportions Graphs

Example question:

"What proportion of the study population was female?"

When your question relates to the proportion of the study population(s), then you will often be best using a stacked column bar chart.

Stacked bar charts and pie charts appear much less frequently in scientific papers, but they might be the simplest/easiest to understand chart type for *your data* so are worth including on this list.

Stacked bar charts

Stacked bar charts are a way of showing the breakdown of your population into subcategories and can be plotted so that each population reaches 100% or so that the bars have different heights to represent absolute values. The second option means you can deliver the total number *and* the breakdown into sub-categories in a single graph, which can be particularly useful if both parts of the story are important. As usual, a benefit comes with a compromise, in this case the absolute graph might make it harder for a reader to spot more subtle differences in subpopulation proportion.

Stacked columns can work well, but they have two notable limitations:

1. When you have many sub-categories, stacked columns are hard to interpret. **Only use a stacked bar chart if you have five or fewer sub-categories**.

2. Error bars are difficult to interpret in these types of graph; therefore, it is hard to show measurement variabilities within populations.

Because of the problem with displaying error bars, most of the time, you will end up showing a dot plot of the different ratios when you go to publish.

The stacked column might be useful for a poster or talk as it can deliver the message more effectively (or consider using the stacked column in your primary figure and include the dot plots in a supplemental figure).

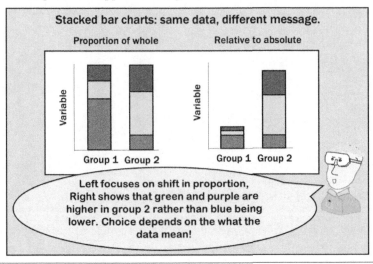

Stacked bar charts: same data, different message.

Proportion of whole / Relative to absolute

Left focuses on shift in proportion, Right shows that green and purple are higher in group 2 rather than blue being lower. Choice depends on the what the data mean!

Pie Charts

Pie charts are not used very often in primary data reporting for the simple reason that it is very difficult for a casual observer to interpret pie charts accurately. The one common exception is in -omics data when talking about broad proportions. This reflects their main use; pie charts are best when you want to emphasise that one or two sub-categories make up the bulk of the population. They aren't really about the precise numbers, more the *general trend*. For this reason, you see pie charts in popular science pieces and on news reports a lot more often and you could consider them if you want to make a visual abstract, write a public engagement piece or for talks or posters.

If you decide that pies are the answer, be aware that rrror and uncertainty are even harder to illustrate in a pie chart than in a stacked column. Indeed, it is almost not worth trying. Instead, you could/should write the % on the chart along with the relevant dispersion.

As for stacked columns, pie charts work best with low numbers of subcategories probably in the two to six category range.

Time-to-an-event plots

Example question

"Do patients with high levels of X expression develop metastases early"

To answer this question you would need a time vs proportion of the population with metases plot.

Kaplan Meier survival plots

The final group of common graphs are for when you are plotting or comparing the time until *something* happens. A binary outcome relative to a continuous independent variable. These are commonly used for describing patient data with the "event" being death of the patient or perhaps relapse or recurrence of the disease. The time axis could also be measured in something like "number of cycles" when doing a time-to-break type analysis of a new material. These sorts of graph are often called *Kaplan-Meier survival* curves.

In Kaplan-Meier curves, there are two potential events at each time point. They can experience the *event* (a 1 is entered at the time point the event occurred), or they can withdraw from the study, meaning no more data will be available. These withdrawals are termed *right-censored* or just *censored*. Traditionally, censored data are entered as "0" into your stats program and are denoted by the upward ticks on the curve..

In terms of data analysis, you are likely to use a form of *logistic regression*, the *log-rank test* (Mantel-Cox) test to determine if the difference between the curves could have occurred due to sampling. This is based on the assumptions of the Kaplan-Meier analysis method, which makes the prediction that the events should happen equally frequently in the comparison groups. A variant of this test is the *Gehan-Breslow-*

Wilcoxon test which puts extra weighting on the early time points. Again, this is something to look up if it becomes relevant for your data.

The stats test you perform on these types of data, as usual, tell you whether the *difference* you have observed is likely to be a false positive. However, the test does not tell you anything about the magnitude of the effect, i.e. whether the observed difference is biologically important. A common method for determining hazard ratio is the *Cox proportional hazards model*. As usual, your stats program will be able to calculate all these for you, so look up the method online, and you will be all set. Usually, it is this *hazard ratio* that is the most interesting to a reader and therefore, it is better to emphasise this aspect of the results when writing your results. "Patients with the high levels of X expression were 3.2 times more likely to develop metastases than those with low expression."

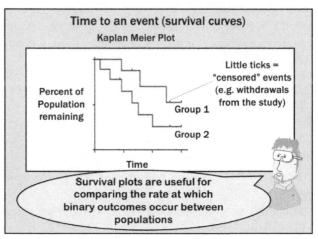

Still unsure which type of graph to use? Clues from stats

An alternative way to consider which graph to display, is to consider the stats you're your performed;

Statistical Test	Option 1	Option 2	Option 3
Chi-squared	Stacked column	Bar chart	Pie chart
T-test	Dot plot	Box and whisker	Overlaid histograms
1-way ANOVA	Dot plot	Box and whisker	Violin plots
2-way ANOVA	Dot plot	Box and whisker	Violin plots
Repeated measures ANOVA	Line graph		
Linear regression	Scatter plot		
Log Rank Test	Kaplan Meir survival		

Appearance: making your graphs publication ready

Ok, you've tested out a few graphing options and decided on the one that conveys your data in the clearest most effective way. Now, all you need to do is format your graph so that it looks professional. Before you begin, have a look several published papers to get a feel for what you are aiming for. Make sure to look in the better-quality journals, as you want your work to be of a high standard. You'll quickly notice that most graphs have a distinct look to them (and that look is *not* the default red/blue version generated by Excel).

Graphing software

There are lots of choices of programs you can use, and it really depends on how much you want to pay. Excel and Numbers are logical places to start as you almost certainly have access to them but don't use the default formatting for your graphs. Take the extra 5 mins needed to make your graph look like better. Both programs are fine for line, scatter and stacked column charts but not so easy to use for dot plots and box and whiskers (it is possible, but it is more hassle).

Stats programs like SPSS, MATLAB or Minitab, all have graphing options. They take a bit more getting used to either in terms of learning the commands or finding the tools to get the formatting right but they can give you much more impressive outputs. As mentioned for the stats part, R is a powerful programming language with lots of free packages available to make glorious graphs. A good place to start is ggplot2, but there are loads of other options! If you are starting on a career in science, it is probably worth taking the time to learn this language.

There are also dedicated graphing software such as GraphPad Prism, Origin and Sigmaplot. Some of these are quite pricey but what you get for the money is quicker or more intuitive platforms that generate graphs that are closer to being publication-ready and which are easier to manipulate. If you are starting out on a PhD pathway, it might be worth investing in one of these. There are free trials available for most, so you give them a road-test first.

Consistency

Whatever program you are using, be *consistent* with your graph formatting across *all* your figures. Use the same fonts, font sizes and colour schemes. Also aim to keep things like axis colour and widths, gaps between groups, error bar formatting and even the lengths of the X and Y axis consistent. For most graphing software, you can save templates to save time and to help maintain graph-to-graph consistency.

Big Tip

Set up templates for your commonly used graphing approaches to help stay consistent.

Colour schemes and point markers

Remember, it's all about data delivery; everything should be as clear as possible. In theory, you can choose whatever colour scheme you want but, in addition to the consistency comment above, I have two suggestions here. First, don't use the default blue/red colour scheme of Excel. This isn't because it is particularly bad *per se* but rather that it says that you haven't spent much time thinking about presentation. Just a couple of quick clicks and the negative connotations will be avoided. Second, choose your colour scheme to help your reader rather than at random. I suggest putting all your control treatments in black/dark greys and all your samples in white/lighter greys (or vice versa). This makes it easy for your reader ti immediately know what column to compare with what, which will help them progress through your manuscript. If you can use a colour scheme that means something (such as your green fluorescent protein samples having a green fill), you should do so, but don't use overly bright colours for no added benefit and keep the number of colours down, grayscale is almost always totally fine. Don't forget to think about colour blind people. Point markers should be big enough to see, which means having solid fill in a dark colour.

Fonts

As with the rest of your figures, there are set fonts that each journal will allow. Arial or Times are generally accepted and so that's what I recommend here and definitely avoid the Excel default Calibri. There are also minimum size fonts (6 or 8 pt. is usual). These font sizes refer to the text *after* the figure it is fully assembled. Therefore, if your graph will make up one panel of larger multi-panel figure, then make sure the graphs will meet the requirements after resizing. Also, remember that the fonts on other labels on your figure need to be consistent with the "maximum two font sizes per figure" rule.

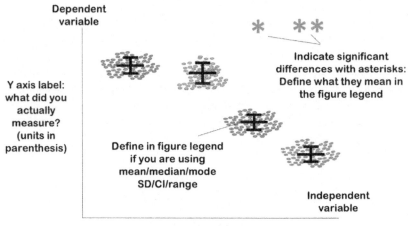

Title: what was the assay

Labels

Your reader should be able to fully understand what you have graphed without having to read the legend. If you haven't guessed already, I am a fan of more labelling, including a title. Make life easy for your reader and they will enjoy reading your paper more.

Make sure your X and Y axis are fully and appropriately labelled. The X-axis will almost always have an independent variable on it; instead of 1, 2, 3, or any other meaningless label, try to use actual words or some sort of acronym that will identify the population (and define the acronym in the legend). For the Y-axis, the biggest problem is usually not enough information being presented. If your measurements have units, make sure to indicate them on the axis. If you have plotted something like "relative expression" you should be indicating what it is relative to. For indirect measurements, the axis should say what you have actually measured and not what you interpret from those numbers.

Size

Probably smaller than you think. Have a look at some published papers, and you will quickly realise that the graphs aren't huge, and nor do they need to be. You don't need a whole page width four-column graph.

For most standard graphs (scatter, box and whiskers, etc.), a height of 35 mm and width of 9 mm x the number of columns is a good place to start. If you have more complex data, many groups, etc., you can justify using slightly larger.

3D

Nope. Don't use 3D graphics including shadows *unless your data has three dimensions* (a principal component analysis or three-axis plot). Keep your graph as clean and simple.

5.4 Figure Legends / Captions

You've made your figure, now is the time to write the legend or caption. Easy, right? Well yes and no. Somewhat surprisingly, figure legends are an area where lots of writers miss the mark or struggle at first. However, once you have done a few, you end up banging them out quickly without any issue.

Figure legends exist to tell the reader what they need to know to *understand and interpret* the figure. The challenge is identifying what needs to go into the legend versus what is covered in the results and methods sections.

What goes into a figure legend?

Journal Rules

Most figure legend content rules are consistent from journal to journal. The only differences I have observed are that some are a little more definitive in their instruction and leave less room for interpretation. The phrasing below comes directly from the journals. You will all these details in a page labelled "instructions for authors" (or equivalent) on the journal website.

- **Nature:** Each figure legend should begin with a brief title for the whole figure and continue with a short description of each panel and the symbols used. All error bars must be defined.
- **Journal of Cell Science:** The first sentence of the legend should summarise the figure and be in bold. Each figure legend should stand alone and should contain enough information to ensure that the figure is understandable without having to refer to the main text. Figure panels should be labelled with uppercase letters (A, B, C, etc.), and each panel should be described in the legend. Any abbreviations not given in the main text should be defined.
- **Journal of Biological Chemistry:** Legends should contain sufficient detail to make the figure easily understood. All symbols should be defined and all equations used to plot lines and curves should be provided. Legends should contain all appropriate permissions statements for reuse of content from previous publications.
- **Scientific Reports:** Figure legends begin with a brief title sentence for the whole figure and continue with a short description of what is shown in each panel in sequence and the symbols used; methodological details should be minimised as much as possible.

In addition to the content, journals will define where in the document to place the legend, whether or not a title is allowed and may put limits on length.

Examples of journal-specific differences in figure legend rules

Journal	Location	Length	Title
Nature	Initial submission in doc Final submission at end	100 words	Yes
J Cell Science	End of manuscript	N/A	Yes
J Biol Chem	Underneath Figures	N/A	No
Scientific Reports	End of manuscript	350 words	Yes

Now we know the rules, let's look at the execution. As with the other sections, using published examples will always help. Some of the following is a little subjective; different supervisors will have opinions about how much or how little to include. Therefore, start by looking at your supervisors' most recent papers and study the style that they use. This is especially worthwhile if you can find a paper in which they present similar types of data as you are.

Big Tip
Read your supervisor's recent papers to find out what style of figure legend they prefer.

Title statement (one clause, usually in bold typeface)

The figure title is either a *descriptive overview* of the whole figure or a *declarative statement* of what the figure collectively shows. These two options are the opposite of each other. My suggestion is that for most figures you opt for the declarative version i.e. say what you want the reader to *interpret* from the figure (see examples below). The reason for this is simple, all other aspects of the figure legend are supposed to be descriptive, without any conclusions. Therefore, take the one opportunity you have to make a statement. The exception of course is for figures that are about setting up a model or validating an experimental system. Descriptive are fine for these.

The figure title should be short and to the point. Most journals limit titles to one line of text.

Example figure titles

Real examples from Journal of Biological Chemistry papers:

- March-I E3 ligase activity is not required for its ubiquitination.
- Bovine PERK luminal domain (bPERK-LD) can directly interact with the denatured model proteins and suppress the heat or chemical induced protein aggregations.
- NANOG antagonize OTX2 to regulate neural patterning in hESCs.

Notice how that each of these figure titles sound like they could be the title for a whole paper. They tell you the *answer* rather than the *question*.

Description of how the data was acquired

This is the bit that new writers usually miss. For each panel of your figure, first describe how the data was generated or where it came from. Not too much is needed here, this isn't the methods section, the trick is to write just enough to aid the interpretation of the images/graphs/etc. If you have multiple panels generated in very similar experiments, then use a general sentence describing those multiple parts rather than repeating yourself.

The details you need in your figure legend are **only** those that are relevant to the *interpretation* of the data and those which are not obvious from the figure itself. For example, if you are showing pictures of cells that you have processed for indirect immunofluorescence microscopy, then your figure legend might say what cell type, onto what substrate (glass/plastic etc) and for how long, with which antibodies, and whether the images were from epifluorescence, TIRF, or confocal microscopy or whatever. You may not need all these details; it will depend upon how completely you have labelled the figure, and which details are relevant to the interpretation.

The same is true for graphs. It is not enough just to say, "graph showing...." you need to say a little bit about where those numbers came from. What did you measure? How were the data processed?

Big Tip
Your figure & its legend should be able to stand on their own; your readers shouldn't need to read the methods to be able to interpret the findings

Description of what you are showing in the figure

Say what you have presented in **every** panel of your figure, **in order**, e.g. A) "representative images of....", B) "scatter graph of...". Describe what is where; "left and middle panel are single-channel images with antibodies..., right panels are merged images". Describe colour schemes; "antibody X is pseudocoloured green and antibody Y pseudocoloured magenta, with DAPI images in pseudocoloured blue".

In your figure legend, you must describe what you have plotted in your graphs. Have you indicated the mean, median or mode? are error bars SD, 95% CI, or interquartile range? what do your boxes and whiskers represent? Does each point represent a single experimental unit or a summary statistic? Expand any acronyms or shorthand in your labelling. Also remember to indicate what you have normalised to or against (e.g. for ddCt or blot quantification, which reference transcripts or proteins were used). Your reader needs this information to understand the decisions you have made.

Scale bars

If you are using microscope images, make sure the scale bar has been defined. You may have directly labelled the scale bar on the image, in which case you do not need to repeat this information in the figure legend. However, if you added the bar in your imaging program then resized the image when assembling the figure, it is likely that the scale bar label will be too small (below 6 or 8 pt). If this is the case, either relabel the scale bar or remove the on-figure text-label and define the bar in the legend.

In an all print era (i.e. the past) you knew the size your figure would be printed and could state the magnification of the image based on that size. Now, your work will be viewed on a screen of some type and you don't know what size monitor or how zoomed in your reader will be looking at your figure. Therefore, use a distance measurement (μm) rather than absolute magnification (x200).

If all images are the same scale, don't repeat this information. Write it once, usually at the end.

Statistical analysis and population information

Always indicate the number of values you have plotted, the independent experimental repeats you performed to generate your data. This is important; the experimental "N" numbers should be explicit in your legend if it is not already clear from your figure. You can also comment on the number of technical repeats used to generate each independent experiment, but usually, this piece of information can live in the methods section. Report experimental N even if using representative images, "representative images from three experimental repeats".

You should indicate which statistical tests you have performed including post hoc tests, the thresholds you used to reject the null hypothesis (e.g. "* indicates $p < 0.05$"), which groups were compared, and how you accounted for multiple comparisons. More extensive details for all these will be in your materials and methods section but, again, include enough in your legend for the reader to interpret the data without having to hunt for the important pieces of information.

Tables

Tables and figures are numbered independently unless the table is a panel within a figure, i.e. you might have a Figure 1 and a Table 1.

Above the table, you place a single clause title "Table 1 …". The legends (known as footnotes) beneath a table are usually much shorter than a figure legend. Define acronyms and any extra symbols which need to be explained.

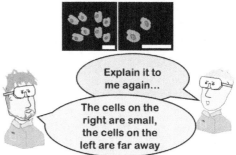

Make sure you define your scale bars in your figure legend

As usual, if something is not clear from just looking at the table, then you should expand a little, but really you should be using headings in a way that makes it clear what is being reported.

What *does not* go into a figure legend

Lots of methods. Ask yourself; is the information needed to *carry out the experiment* or is it needed *to interpret the figure*? If it is the former, then the info goes in the methods.

Results. The job of the figure legend is to describe *the figure*; you don't describe any of the actual findings. There should be no descriptive statistics in the figure legends, those details go in the results section. If you find yourself writing "mean of X…" then stop and delete.

Discussion. Except for the figure title, a figure legend should not include any discussion or interpretation of what the data means or why you have done those experiments. Again, you have other sections of your writing that are specifically for this purpose.

If you feel that you need to add something extra to your figure legend to explain what you are trying to show it is a very clear sign that your figure isn't as good as it should be! In these cases, go and fix the figure rather than trying to correct the problems in the figure legend.

> **Do not repeat between the figure, the figure legend, or the manuscript text any information that appears in one of the other two.**

What to check for when editing

During editing, carefully check that **all** the panels have been described in the order they are presented. Make sure that you have included enough information to allow the reader to interpret everything. Check again that experiment "N" numbers, scale bars and stats are described. Once you have drafted your legend give it and your figure to a friend to make sure it makes sense. Remember, each figure plus its legend should be able to stand alone.

If you have followed all these instructions and written everything, most likely your legend is really long. Too long. Therefore, in the editing stage, you should look for ways in which you can combine and contract information to make the whole legend more efficient. Make sure you don't have results or discussion and that everything else is a succinct as possible.

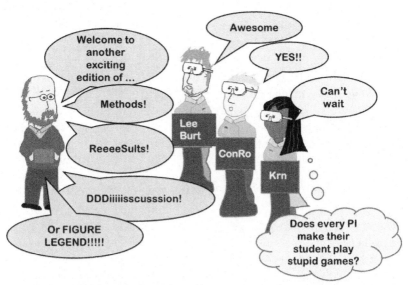

The danger with figure legends is that you repeat things from other sections of your manuscript. Only include things that are needed for readers to interpret the data in the figure. Don't include any results or discussion, and only use the bare minimum of methods-type statements.

Chapter 6: Writing

6.1 Before you begin writing: General Tips
Big concepts and general advice about the writing process.

6.2 Microsoft Word Tips
Useful tools and other tips you might not know for working with manuscripts within MS Word.

6.3 Notes on scientific publishing
An overview of the steps required to get your work published.

6.4 Literature reviews and review articles
How to summarise your field in a way that is useful to your readers.

6.5 Writing for the general public
Tips on how to make complicated science stories accessible to the general public in blog posts and press releases.

6.1 Before you Begin Writing: General Tips

Project report, manuscript, dissertation or research paper?

Throughout this book, I have used the terms report, manuscript and paper almost interchangeably. However, *manuscripts* refer to any unpublished work. The same text becomes a *paper, article, monograph* or *book* once it is accepted and published by a journal or a publisher. Of course, you might also describe the work based on the type of writing; a *review, commentary* or *letter.* Confusingly, a paper could also refer to a presentation at a conference, which are also known as *abstracts*, which are often published in journals too. In addition, in many Universities, the work that students submit are called *papers* (usually USA) or perhaps *project reports* (usually UK). *Dissertations* and *theses* are normally longer pieces of work with more extensive introductions and discussion but could be referred to as manuscripts and could, ultimately be published into a book.

While all that is confusing. The good news is that most of the writing advice presented in this book is equally applicable to all the different forms of writing. The only exceptions are the introduction and discussion sections. Chapter 6 deals with things that are generally true irrespective of which type of writing you are doing. Chapter 7 has specific sections of each of the specific sections of a research article (or that style of paper) as this is the type of writing that is the most important and common mechanism of talking about research outcomes.

Write for the Reader

> **Big Tip**
> What interests you is not necessarily the same as what your readers will value.

Throughout the writing process, you should be considering the needs of those who will read your work. This means that everything should be delivered in a way that makes it as easy and as enjoyable as possible to absorb and digest. However, it's not just the small things like sentence structure that you need to consider, the primary think you should be thinking about is *why* they are reading your work. If you all your writing so far has been a course paper or exam, then the person reading your work has been paid to do so. In the real world, it's the other way around. The reader needs to get something back from the time invested. That means, you have to provide *value.*

This concept of providing value should stay with you irrespective of what you are writing. It should be in your thoughts when you are preparing your figures, when you are writing up your results, and when you are

Effective paper writing will engage the reader while delivering a focused message with a clear narrative

leading your potential readers through a complex discussion of what those data mean. It's the first thing you should think about before you start any planning and the lens through which you should edit your work. How can I deliver value?

To deliver value, you need to know who your readers are. If you are writing a manuscript for submission to a journal, you should consider what the scope of the journal and what that means in terms of its specific readership, what those people are used to seeing, what they are likely to be interested in, and what level of jargon is acceptable. If your work will have multidisciplinary interest, then you will need to write differently to ensure it works for every discipline than when writing within a very narrow niche.

You do not have to try to sound smart

As a new writer, you might feel imposter syndrome, that you need to prove yourself somehow in order to be accepted by experts in your discipline. I would love to be able to tell you that there was an easy way to avoid feeling like that, but I don't have one (yet). However, just because you feel a certain way does not mean you have to act on it.

You **do not** have to use the most verbose, grandiose verbiage in your writing. Each sentence does not need to be 30+ words long with at least four sub-clauses.

Your writing should be easy to read. Use the most precise language for maximum accuracy but do not choose more complex phrasing unnecessarily.

Yes, you should use the correct words, but please do not look up the thesaurus for the most complex synonym just to try and sound smart!

> **The best writing is effective writing.**
> **Writing that can be read once and understood fully.**

Put yourself in "writing-mode."

It is important to acknowledge that some parts of writing are easier than others. For example, you might find the materials and methods section to be quick and easy whereas the discussion is more challenging to get right. Similarly, the quality of your writing or how difficult you find putting together coherent sentences varies depending on your mood, tiredness, time of day, environment etc. Get to know yourself in terms of when you do your best work and then plan your writing periods around those times. Moreover, do what you can to put yourself into the right mood. This means rather than a marathon writing session, build in resets every so often; go for a walk, do some exercise or go do something else to recharge your body. Whatever you need to do your best work. Depending on your course structure, especially for a Doctoral or Masters thesis writing, there might be a a dedicated "write-up" period. During these times of extended writing, the key is responding to your body and working on the right part of your document at the right time.

Personally, I find that my deepest thinking happens about 30 to 60 mins into a writing session. Sometimes, I get into "the zone" beforehand by going for a walk and thinking about the problem without the distraction of a computer in front of me. Other times, when deadlines are looming, I will work on the non-deep-thinking parts of the document to get going and then swing back to the hard stuff once I'm working well.

Writing, or writing well at least, requires concentration. Turning off social media and email alerts while in the heavy thinking parts of a document are also a must. There are some nice apps that can do that for you, have a look for pomodoro times.

Make your writing sessions bite-sized

There are lots of elements to a well-written science paper, and it can be quite intimidating looking at the whole thing as a single entity.

Take breaks, go outside, do some exercise. You need to be mentally alert to be able to write effectively

Once you have identified what each paragraph needs to deliver, it's suddenly a lot less daunting; you can focus on one paragraph or subsection at a time. Moreover, thinking about small parts means you can make meaningful progress in relatively small packets of time. For example, you could write a figure legend while waiting for an incubation step on an experiment to finish. Materials and methods sections and results subsections are great for fitting in around your work as they don't require much dedicated thinking. If you are doing a PhD or Masters, learn how to structure your day so that you are able to do this and you will finish much more rapidly.

Multiple drafts

Big Tip

Get started. Don't worry about finding the perfect wording in your first draft. Focus on getting the bones of the story laid out.

Your first (and second) draft will not be perfect. The good news is that they don't have to be. Editing is an integral part of the writing process, and you should include dedicated time focused on editing in your schedule.

One of the hardest parts of writing is getting started. Once you accept that you are not even aiming for perfection in every sentence in the early versions of a document, it is much easier to get over any initial inertia.

The first draft is all about getting the storyline together. Focus on working on the flow and structure. During this time, I recommend generally either writing *or* reading. Close your browsers/pdfs and write. You can and will be able to add fine details later. Just get *something* onto the page.

Writing is part of the thinking process

Why is it hard to get started? I think that this a product of how we learned to write. When we were in school, our teachers taught us to do all our thinking first, then formulate a writing plan, then, and only then, start writing within this plan. However, this order of working becomes very difficult when the subjects are more complex. The actual process of writing becomes increasingly important as a mechanism for you to digest the complex material and to clarify your thinking. You are likely to find that writing helps to put your ideas in order and to work out what everything means.

Big Tip

If you are writing about a complicated topic, your first draft is likely to be useful for you, but it will be terrible for a reader. Don't be afraid to start again.

Unsurprisingly, the writing you do to develop your thoughts is very unlikely to be the sort of writing that is best for your reader. Acknowledge that when working on a complex body of work, the first things you write are very unlikely to be even close to ready for submission. Editing of this draft isn't a case of word choice tweaks or deleting the occasional sentence, it will involve restructuring large pieces, moving things around and completely removing or adding new sections. In some cases, it might even be more efficient and lead to a better final document if you started again, but this time armed with your new clear direction of where you are going. I've taken to calling this draft my *thinking draft*. It comes before my first draft proper. Writing it isn't a waste of time. To the contrary, it means that the first draft ends up being closer to where it needs to be because you will now know the best way to approach the work.

It takes a level of self-awareness to be able to say, "this draft is just for me, I will start again," but, as soon as you *can* do that, you will be able to write the drafts very quickly, without worrying about spelling, grammar, sentence and paragraph structure or picking out the details. Instead, you can focus on the *thinking* rather than the writing aspects.

Don't expect your first draft to be perfect. The first draft is to get started. Use writing as a way to process your thoughts and make sense of your story.

This will not only generate a document with a structure that flows more effectively for the reader but also will be easier for you to write.

What to look for when editing

In each of the subsections of this book, I have included a section where, in my experience, students or new writers often make mistakes. You will have seen these already in the figure legends part of chapter 5. Some of these parts might sound a little repetitive if read immediately after the section to which they are referring; however, I have added them for two reasons. One, they emphasise the important bits not to forget and two, so that when you are editing your own work, you can quickly flick back to the appropriate section and remind yourself of some of the more common problem areas that you should focus upon.

Chapter 8, later in the book, goes through the stages to editing a document. Editing is important. It takes time and practice.

What's your Story?

Plot a Course to Success

Early in the writing process, you need to decide what *story* you are going to tell. If you are going to take the reader on a journey towards a new discovery or a new way to think about the world, and, most importantly, if you are going to make them care about it, then you will need to have a good idea of where you are going throughout the process.

If you are writing a small, linear study (for example for short-term project reports such as Masters' or undergraduate projects) then you probably already know what the main story is. However, as you progress in your scientific career, you are likely to need to tie together a series of experiments in order to detail a much bigger finding More content means more options as to how to assemble the different parts.

Usually, the key messages you want the reader to take away are the biggest, most exciting, most important findings from your work, but deciding on what are the most important parts is not always easy. It requires you to step back and look at everything from a much wider viewpoint. A good idea at this point is to talk to people about your data. Try different ways of engaging them to see what works best. Don't just do this with people in your own field, think about how you would describe this specific body of work to someone in the pub/elevator/at dinner etc.

Remember, what is most important to you may not be the most important, exciting, or interesting part to someone who has not heard about the data previously. Give the people what they want!

How to build a cohesive story

When you are writing up a primary data scientific manuscript, the results part of the story is dictated by the data. Therefore, deciding upon a main storyline and the making of the data figures go hand in hand. A jig-saw session can really help to play with different narratives.

For literature reviews or lay articles, you could take the same approach. Gather together the pieces of information you want to deliver and have a sort of metaphorical jig-saw session. Play with assembling the pieces of the story in different ways, saying out loud how you would connect from one piece to the next. I find this useful not only for planning the writing of a manuscript but also for identifying where there are holes in the story; i.e. the next experiment or where I need to read more.

In your first draft, rather than trying to write whole paragraphs straight from the beginning, you should set out a framework to focus on how the story flows. Think about what each individual part contributes to the main message you want to deliver. Think about the rationale behind each section; does one set of findings inform the next?

Big Tip
You don't have to present your work in the order you carried it out. Write in an order that helps the narrative.

Don't be afraid to leave things out. As you assemble the pieces of your story, you may find something that doesn't contribute to the core message. I know it's difficult to leave out parts that you have spent time and effort carrying out, or even tranches of the literature that you have spent hours learning about, but if the information is not relevant, then your readers won't value it, and it will become a distraction.

However, I need to draw an important distinction. If the data *are* part of the story, you *cannot* leave them out just because you don't like them or they don't fit the narrative. You are ethically obligated to report everything that is relevant.

Write in Paragraphs, Think in Paragraphs!

If it is a long time since your last English class, you might not have thought about paragraph structure for a while, so here is a quick recap:

No more than one key point per paragraph!

The first sentence of every paragraph (the *topic* sentence) should cover, in relatively simple terms, the main point you want to deliver. These sentences are really important. They should be able to advance the story entirely on their own. Indeed, people who are skim-reading your work will use the topic sentence to decide whether they need to read the rest of a paragraph or not. You should aim for your topic sentences to be accessible, quite general and fairly short. The more complicated, more specific and deeper materials come later in the body of the paragraph.

Paragraph Structure
Topic
Token
Link
Token
Wrap

When you sit down to write a new piece of work, often you will be able to identify the main point you need to deliver quite easily, but the details might take you longer and might require reading additional papers. You can use the simplicity of topic sentences to establish an outline for your paper quickly, and this can allow you to focus on the parts that you are ready to write. With the outline in place, you can come back later to focus on fleshing out the deeper details of each paragraph once you have done any additional reading or thinking.

The meat of your paragraphs are the *token* sentences. These contain the detailed information needed to support the topic sentence. The token sentences are where you provide the evidence or the specifics and they will always require citations, usually to the primary source of the information (more on referencing shortly).

> **Big Tip**
>
> Structure rules help you avoid mistakes, but you don't have to follow them rigidly. Occasional breaks from the standard structure can be used to increase impact.

Token sentences tend to be information-dense, so if you have a stretch of multiple tokens back to back, it can be hard for a reader to digest. To keep your readers happy, set yourself a rule of no more than two token sentences in a row. Use softer *link* sentences (or link clauses) to break up stretches of token sentences and soften the delivery of heavy material. Link sentences can also be useful to reconnect the whole paragraph back to the core message you established in the topic, or to pivot slightly to sub-point. Remember, only one key point per paragraph.

Finally, *wrap* it all up. Finish your paragraph with a sentence that closes out the paragraph and leads you to the next one. Having a wrap sentence will ensure your reader knows exactly what they are supposed to take away from a paragraph and will ensure the whole piece has a pleasing flow. Wraps are your way of linking from one paragraph to the next and they make a huge difference in terms of establishing the flow of a piece.

A good paragraph is like a burger. The bun is the topic and wrap sentences that give the reader something to hold on to. The meat are the hard facts, the token sentences. If you have a lot of meat? You'll need condiments (links) to stop it from being too dry.

Check the instructions / syllabus

I have tried to make the information in this book appropriate for as many different styles of assignment or journal articles as possible. However, it is impossible to cover every single scenario. Every journal has specific rules and, of course, every University assignment has its own rules and marks schemes.

> **Big Tip**
>
> Save yourself some work: check the "instructions for authors" or course handbook before you begin.

Identify subsections, word counts, and figure limits

The order, length, layout and naming of specific subsections is different between journals and assignments. You will usually write your first main draft with a target journal in mind or will be writing a project report for your course, so check your instructions and save yourself some work.

Figure limits (in terms of how many you are allowed) and abstract word or character count are two things that can vary quite dramatically between different journals, so it is always a good idea to check these numbers. Some other things to look for are listed below:

Things to check in the "Instructions for Authors."

Common differences between journal requirements to be aware of:

- **Whole Document:** word/character limits, figure number limits, subsection order and names, acceptable file types, line numbers.
- **Title:** Identify word/character limits. Identify other components of title page including keywords, abbreviations, acknowledgements and author list formatting.
- **Abstract:** determine word/character limits and if it should be structured or unstructured.
- **Lay abstract and / or key points list:** may be required for some journals.
- **Introduction:** usually pretty standard between journals but may have separate literature review and introduction sections in longer format writing. May ask for separate hypothesis or aims statement.
- **Materials and Methods:** usually pretty standard but check location as they may come before or after results section. Might have a different name such as Methods or Experimental Procedures
- **Results:** check whether sub-headings are allowed, how figures are referred to in text (Fig. 1, Figure 1, F1, figure 1), some journals combine results and discussion.
- **Discussion:** check if a separate conclusions section is required.
- **References and in-text citations:** check journal style (download to reference manager).
- **Figure Legends:** check if uploaded with figure files or as part of manuscript (usually the latter). Check whether at end of document or in the location of figure. Determine if figure titles are allowed and any word limits.
- **Figures and tables:** check page sizes, fonts and file types allowed, whether locants are A, a, i etc.,

---

Make or download a template

I never write my papers in order from start to finish, and I don't think you should either. One of the first steps I take when I am about to write a new manuscript, or indeed any lengthy piece of work, is laying out a document with all the required subheadings so that I can work on any part at any time based on my mood and preparedness.

Some journals have produced ready to download pre-formatted Microsoft Word and LaTeX templates that can be used. I have made a quite generic .doc template available that you can download and use, just check it fits any specific requirements of your assignment or target journal http://lantsandlaminins.com/writing-guides/manuscript-or-report-template/

UK or US English?

Simple rule here; be consistent. Most journals will accept manuscripts written in either form of the English language, so use whichever version you know best. Whatever you decide, stick with it. Surprisingly, identifying which version of English I am using is one of the hardest things I found in moving from the UK to US and then back to the UK again!

Stay consistent with your language choice

Be aware, some words have different meanings or are used differently depending on the style of English used. When you are aware of any differences of this type, aim to use a synonym instead to avoid any ambiguity.

Look at examples

I will keep coming back to this point. There is wealth of published scientific literature out there, so use it. You do not need to struggle coming up with a brand-new way to say something, use a variation of what has gone before. For each section of your paper, have a look at how similar types of writing have presented their findings. Look at the phrasing, look at how the data figures and tables are presented and described.

There is an important distinction between learning from what has gone before and plagiarism. You can't directly lift passages. However, you can learn from what works.

You will quickly realise that some papers are better than others not only in terms of the quality of the data or its impact but also in the way that the work is presented and how much you enjoy reading it. Try to identify what it is that makes paper easier to absorb and adopt those aspects of style. Likely, you won't notice these things until you look for them, so build a bank of papers that you thought "worked" then re-read those looking specifically at the style rather than the content.

6.2 Microsoft Word Tips

Other manuscript preparation software is available, including LaTeX and other free programs. However, as many people use Microsoft Word, this section contains some handy tips that I find useful in my writing.

Useful Commands

Quick Selections

Left-click once to select a character.

Click twice to select a whole word.

Click three times to select a whole paragraph.

Press Ctrl+A (Cmd+A on a Mac) to select all.

Jump to specific location on a blank page by double-clicking on that location (line breaks and spaces will be automatically inserted into the document to allow your cursor to arrive at the appropriate place).

Formula and chemical symbols

Press Ctrl and + to write in superscript

Press Ctrl and − for subscript.

When using Greek symbols, it is a pain to keep changing the font to symbol, and can also run a risk of it converting back to plain text if you later change the font of the whole document. Your μm turning into mm is never a good thing! Instead, use the insert symbol command. However, that is a slower process and not something you want to have to do often. Instead, set up shortcuts. 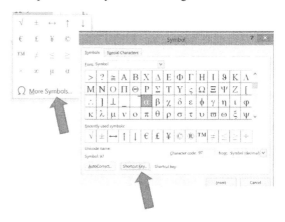 From the *insert symbols* command, click "more symbols", choose the symbol you need and select *shortcut keys*, then choose something obvious that will work for you and is not already assigned to a command you use. For example, I have set up Word so that Alt+A is α, Alt+B is β, Alt+g is γ and Alt+m is μ.

Move around faster

If you are working on a lengthy document, you can press Shift+F5 to cycle through each of the different points you edited most recently. Similarly, if you open a saved document, you can click Shift+F5 to take you to the location where you were last working.

Change the sentence case

Each sentence can be written in three options: UPPERCASE, lowercase or Camel case (first letter is a capital). If you accidentally end up with the wrong one or if you later change your mind about how you want your headings to look, you can switch between the three using Shift+F3.

Styles and Tables of Contents

The one tool that makes a huge difference in terms of time saved and consistency in longer format is the *Styles* ribbon.

The styles options will be on your home tab. Each paragraph of text is assigned to a specific style, which you can define. Why is that useful? Well, quite simply, it turns all your formatting changes into single button presses and ensures that every paragraph of a single type looks the same wherever it is in the document.

As you can see from the screengrab of my Word, I have set up Styles buttons for each of the different repeating element used in this book. All of the 393 times I have used heading level 4 have the same font, font size, bolding, line spacing before and after, all loaded with one click. If I modify the style, every incidence that that style has been used will change too.

Word comes with a load of style galleries pre-loaded so if you do not have any desire to become a typesetter you can skip the thinking and just select whatever version works for you. Conversely, when you are working to the specification of a specific journal, you can modify each style so that fits the requirements and also save that style catalogue for future use.

Tables of contents

Using "styles" really comes to the fore is when you need a table of contents (TOC) such as for a thesis or dissertation. Click on the *review* tab, then *table of contents* and select the appearance that you want. Using the default options, any time you have used the style *Heading 1* will be presented with a defined level of indentation, and *Heading 2* will receive the next one down etc. You can click on "custom table" to further refine which styles get indexed in the TOC and how they will look. Once your TOC is inserted, if you then edit the document click

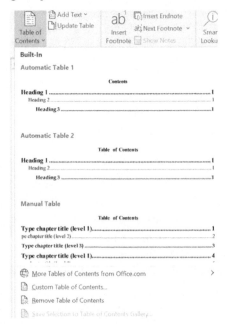

on the table and select *update field* to update the appearance, line numbers etc.

Styles also help in navigation. Once you start using styles, you will also see the

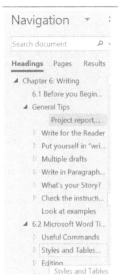

headings appear in the navigation pane. Clicking on those headings allows you to rapidly jump around within your document. No more scrolling through hundreds of pages! When your PhD thesis hits 300 pages, you're going to be very pleased that you used this simple feature.

Editing

Track changes

Most documents you write will have multiple authors, each of whom will contribute to the writing process. Indeed, that is a prerequisite of authorship. Every person involved will have their own opinions and will make edits to the document. Knowing who wants what, and what they have changed is very useful. On the Review tab, you will find *track changes*. Before sending out a document, you can click this on to encourage others to use it (you can even set it to lock-on to ensure that they do).

With track changes on, you have options as to how the document is displayed; *Simple Markup*, *All Markup*, *No Markup* or *Original*. Each of these has their uses, so switch between what works for you. I find it is easiest to spot small mistakes in spacing or alignment with No Markup, so that is how I usually work. I use All Markup whenever I get back minor edits on a late draft. Simple Markup allows any comments to be visible as you work.

When it comes to printing or submitting your final document do not forget to turn track changes off, switch to No Markup, accept the changes and submit a clean version otherwise your recipients will see every step along the way.

Clouds and Comparing version

It is now possible for different people to work live on the same document by using cloud-based storage (integrated in Microsoft teams, or via google docs, dropbox etc). However, if the various authors are each working independently offline on separate versions then another useful tool in editing, is "Compare", which is also located on the Review tab. This allows side-by-side comparison between different versions of a document and enables you to merge documents or select the preferred version. If your co-authors have forgotten to use track changes, then the compare option can help you to spot edits.

Spike: The Extended Clipboard

During editing, you might decide to move around blocks of text to different locations. You can do this one at a time, but another option is to use the *Spike*. This is an extended clipboard where you can add multiple blocks of texts or images one by one to the same clipboard and then paste them all at once to a different location.

To use Spike, select the text or images that you want and press Ctrl+F3, then select the next block and, again, press Ctrl+F3. Once you have added everything you need, press Ctrl+Shift+F3 to paste the content to its new location.

Version control

While talking about editing, it would be remiss of me not to mention version control again. You should save lots of copies, on hard drive and on the cloud or other back up media. However, by playing safe and having many copies, you face the challenge of identifying which version is the most up to date. Therefore, adopt a naming system that will make this clear from the start.

The most straightforward naming system is numeric; working upward from v0.1 (draft) until you reach a threshold where you switch to 1.0 (publication/submission).

Including the date in the name might help too.

Avoid using file names that will be commonly found on your or other people's drives, instead use something specific that will be able to be easily located and identified in the future. As a supervisor, I received hundreds of files saved as "results", "intro" or "report".

Your manuscript will go through many rounds of editing. Use a naming system that will allow you to easily identify which version is the most recent.

Avoid using "final" in the file name, regardless of how hopeful that sounds!

Highlighting

One tool I use frequently is text highlighting. To produce my best writing, I like to stay in the flow, maintaining the thought processes until I get the ideas down on the page. Rather than stop to add some details that I would need to search for, I instead make a note for myself. For example, anywhere where I will need to add a reference I'll put (ref) or (smith et al), and anywhere where the actual data numbers need to be inserted, I'll put (data). In early drafts of a manuscript, I also like to highlight anywhere where figures are referred to (Figure 1A), I do this to make sure that every figure and every panel within each figure has been referred to in the order they appear. The coloured highlights help me spot where to come back to, areas that still need attention, and unlike comments added using the review tab, the coloured highlight is visible when set to No Markup, so it is less likely to be missed.

I also use highlights when doing high-level edits. If I think the document isn't flowing well, then I use the different colour options to pull out different phrases that deliver related material and then reorder those points into new paragraph structures. For example, in a recent manuscript draft we weren't happy with the direction of the discussion, so we highlighted all the phenotype phrasing in yellow, the mechanistic statements in green and the caveats and limitations in blue, then reordered the sentences based on those colours so that we had all the yellow together, all the green, then all the blue. It worked, the new draft had more logical flow. Doing something similar might work for you too if you are struggling.

Headers, Footers, and Section Breaks

Adding a simple header or footer and page numbers to a document is easy (*insert/header footer*). However, in longer-format writing, you might want the header to change as you progress through the material. For example, it could help readers navigate your thesis if each chapter had a distinct header. To do this, you need to use section breaks.

Whenever you start a new chapter (or *section*), instead of using the return key or, inserting a page break to start the new page, use a *section break* instead. Now, when you insert your header or footer, you can have a distinct header for each section. Make sure to turn off *link to previous* (see screengrab) to make this work.

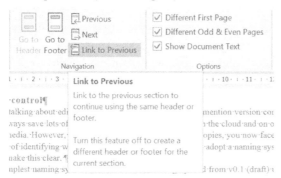

Other tools

Learn how to integrate your reference manager program into word so that you can cite while you write. Other things like the spelling and grammar checkers and the thesaurus are improving all the time and are worth using. At first, they will be frustrating as every science word will be flagged as being misspelt, but if you keep adding to the dictionary, then they become more and more useful.

I do not want this section to get too long with things you already know. Each new edition of word adds new functionality. If there is something you need to do, there is likely already a tool for that. Use the search function within Word, and you will find what you are looking for. As with every program you use, the more you use it, the better you will become, and the more efficient you will be.

6.3 Notes on Scientific Publishing

If you are writing a manuscript that you aim to submit for publication in a journal, then it is worth having an idea of what the process will entail.

Styles of paper

The four main types of journal article that you might read or write:

Full-length research articles (papers).

The most common method used for communicating primary and secondary research outcomes are research papers. They vary quite a lot from field to field and journal to journal but generally about four to six main data figures (with multiple panels per figure), and the text is broken into subsections of: abstract, introduction, methods, results, and discussion, although the order and naming of sections might differ slightly. Many PhD programs will require that you write a full-length paper to graduate and research papers are also the types of writing required for most Masters or undergraduate research project write-ups.

If you aim to have an academic career, primary data research articles are the most important part of your CV. It should not just be about numbers though; your career will benefit more from good quality papers published in better quality journals and those which make an impact on your field and receive citations.

Because papers are so important and as they really are your target communication method for all research data, I have broken down the standard components of a paper and describe how to write each of these section by section in the next chapter.

"Letters" to the editor, brief communications, or rapid communications.

As their name suggests, these are shorter reports, usually with just one or two figures of data and with a restricted word count (around ~1500 words). In terms of structure, they usually contain the same elements as a longer article but are likely to be written in continuous text without sub-heading breaks between sections. Letters can be highly effective to getting your story out rapidly, i.e. without having to wait for the rest of the story to be completed. Therefore, if you have a relatively compact, important and time-sensitive story, it can be worth considering publishing it as a letter. Letters reporting data go through the standard review process of being sent for peer-review.

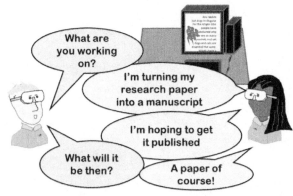

Publication of your work in an academic journal is the number 1 goal for most projects.

Letters can also be used to comment or respond to published work. If you found something where the interpretation was flawed or where the authors failed to consider some other important work, you may choose to respond by writing a letter. You might also be specifically asked to comment on or a paper or series of papers that are coming out. These sorts of interaction will more often come from the well-established professors but, if you think there is scope for something of this type, it can be worth having a chat about it with your supervisor or reaching out to the editor.

Case studies.

There are exactly what they sound like. Case studies report the findings of a single or limited number of specific cases. This form of writing is quite common within the medical and veterinary professions as a way of delivering an interesting or surprising finding and thereby adding to the body of knowledge about a specific condition. Case reports are also a way of describing effective or ineffective intervention attempts, paving the way toward a clinical trial. Formatting is often similar to a research paper, although case reports may not have distinct results and methods sections and likely will have very short discussion (there is not usually as much to discuss).

Review articles, editorials, commentaries or opinion pieces.

Reviews usually do not contain primary data; instead, they pull together the findings from lots of published work and present them as a more comprehensive whole, with some added insight. Usually, but not always, review articles will give deeper and more up-to-date information than is available in a textbook. If you are new to a field, it is usually a good idea to read a few review articles from different authors to get an overview of the overriding thoughts. Review articles will also help you to identify which papers to look for and therefore, which ones to cite in terms of the key points in your introduction and discussion.

> **Big Tip**
> Written a good literature review? Consider sharing it with the field and enhancing your CV by converting it into a review article.

We will talk about literature reviews in the next section. Fundamentally, a literature review could be the basis of a solid review article, and many PhD students turn the first chapters of their thesis into a publication or conversely use parts of a review article they have written as the basis of the lit review for their thesis. Before you begin collecting data, you might be asked to write a review article by your boss as a road toward your first publication.

Some journals allow you to submit review articles for consideration at any time, while others are by invite only. Either way, if you are writing for one of the more reputable journals, your article will be sent for peer review in much the same way as the other writing types.

Steps to Publication: Review and editing

1. Identify an appropriate home for your work

There are literally thousands of journals each with their own scope and acceptance criteria and with different prestige. The decision about where to publish involves considering who is likely to want to read your paper, who will *value* the work, and then looking to see where similar types of paper are published. Every journal has a page describing who the journal is aimed at, what type of data are appropriate, and the kinds of article it publishes. Check these *scope* statements carefully to make sure they fit the story you want to tell. From here on, I will describe this as your *target journal*.

> **Big Tip**
>
> All journals have a page describing the "scope" of the journal. If your article does not fit these criteria it will not be considered for publication.

To write the manuscript well, you need to know who it is for! Ideally, you should decide on your (first) target journal before you write the manuscript, but at very least you should have narrowed your target down to a small number of options. In an all-digital, open-access age where every indexed article is findable using search engines, *where* you publish is becoming less important. However, having a target journal allows you to write your manuscript fitting their stylistic guidelines and tailoring the message to fit the remit of the journal.

The final decision is something that should be made with your co-authors, and a reasonably complete draft or at least the main figures are likely to be required to guide this discussion.

2. Adjust formatting to match your target journal.

Each journal has its own style rules about how their papers are constructed. Before you can submit your manuscript, you need to check these rules carefully; otherwise, your manuscript will get bounced back to you without even making it to the editor. You'll find the rules on the *instructions to authors* page of the journal's webpages. Although each journal does differ slightly, the main elements are usually quite consistent. The biggest differences are in abstract length, figure sizes, and order of the main elements. This means it is possible to prepare almost generic manuscript that will work for most journal but will require small adjustment to finalise.

Once you have your manuscript written, edited *and approved by your co-authors* you're ready to upload. Before your paper even goes out for review, you need to complete one more step; convince the editor that they should review your manuscript. To help convince the editors to make that decision, you will write a cover letter.

3. Write the paper!

No surprise, this is the main part. We'll come back to this in the next chapter.

4. Write a cover letter.

Cover letters are read by the editor and may also be sent to the reviewers. They accompany the manuscript and are usually the first things read. In this short letter, you have a chance to highlight why you think this manuscript would be suitable for publication in your target journal. You should briefly point out what makes your findings valuable, interesting and exciting to the readership of the journal. Do this as concisely as possible, your cover letter needs to be short and punchy enough to get attention but the main story lies in the abstract itself.

The prestige of a journal is driven by the quality of the work that it publishes. The most commonly used way of judging a journal is impact factor which ranks journals based on the mean number of citations that each paper garners in a five-year period. Therefore, in your cover letter, it can help to draw attention to material that you think is likely to be cited.

My approach to cover letters is to use three sections. I start with one sentence opening paragraph stating the title. Next I have a middle paragraph summarising the key findings, highlighting the value* and how the work advances the field. Finally I close with a short paragraph explicitly stating how your manuscript fits the scope of the journal and how it will appeal to the readership. That's it.

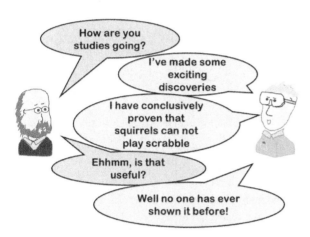

How are you studies going?

I've made some exciting discoveries

I have conclusively proven that squirrels can not play scrabble

Ehhmm, is that useful?

Well no one has ever shown it before!

Just because something is new doesn't mean it has value. Focus your message on why the findings matter rather than emphasising novelty for novelty's sake.

*Be aware that novelty does not automatically equate to value. So, even if you write "for the first time ever", you also need to also say ", and this means..."

5. Upload your manuscript.

Sending your work to the journal used to involve printing lots of copies of your manuscript and mailing them to the journal offices. Thankfully, those days are over; now you upload everything through the journal websites. The processes are getting better, but it still takes a frustratingly long time to type in lots of author addresses, upload the individual files and check that everything has gone in correctly. It is a tortuous process. However, I always feel happy on the day I hit "submit" as the weeks and months working in the lab and on the figures and text are now complete, at least until the reviews come in. A few points worth knowing here:

- For most journals, figures are uploaded separately from the text; there is no need to embed them within your manuscript file. Indeed, you don't want to lose any figure quality in the upload, the preferred option is to upload tiffs of pdfs that you have checked are not pixelated or otherwise lacking in quality.
- Make sure you have all the author names including middle initials and contact details correct. You will be asked for these.
- During the upload process, you will be asked to suggest some reviewers of your work, discuss this with your supervisor to determine who they think

would be fair and appropriate. The editors may or may not take these suggestions on board.

- In bigger journals, you will probably also be able to suggest which editor should process your work. Look on the journal website to find their list of editors and try to identify people working in your field who are likely to appreciate the importance of your work.

6. Editor decides; send for review... or not.

Once the article is uploaded, the editor will read your cover letter and abstract, plus they might have a quick look at your figures to gauge their quality. They will use this info to decide whether to send for review *or not*. If they think it might be good enough to publish in that journal, they will select two to four faculty-level academics with similar research interests to you; i.e. your peers, to review the manuscript (hence the term *peer review*).

Remember that the editors of most journals are academics as their primary profession. They likely don't get paid by the journal. This means that they have their own research and teaching activities that are a bigger focus of

Submitting the manuscript isn't the end of the process. However, you always feel a sense of relief and satisfaction knowing your work is ready to be sent for review

their time and they also might be dealing with hundreds, or even thousands, of papers each year. They are unlikely to spend a large amount of time reading your work before they decide. You need your cover letter and abstract to be very effective to get past this first triage stage.

If the editor decides to reject your manuscript at this stage, then it will not go for peer-review but you will hear very quickly, often within a couple of days.

The reasons for rejection are usually that your work hasn't advanced the field far enough for your target journal or that it is beyond the scope of the journal. If you do get rejected at this stage, take on board what the editor has said and decide if you should look for a different target journal that is better fit in terms of scope or, perhaps less prestigious. Rejection like this is common. In top journals more manuscripts get rejected at this point in the process rather than after review. Rejection does not automatically mean that your work isn't good! Do not be put off. We all want to get our work published in best quality journal we can, so aiming high is not uncommon.

One word of caution, if you know that your work isn't of the quality required for a specific journal you are just wasting your and the editor's time in trying it.

7. Reviewers comments and editorial decision.

If your paper gets sent for review, then two or more experts in the field will carefully read your whole manuscript and decide about its suitability for publication. They will also look for ways in which the manuscript could be improved and suggest these modifications. The modifications could involve editing the text content, or they may suggest new sets of experiments or additions to existing experiments.

Big Tip

Keep your reviewers and the editor happy. Be courteous and respectful in all correspondence.

Reviewing papers is an integral part of being a member of the academic community. It is how we, as a profession, ensure that there is quality in the published work. However, the reviewers of your manuscript are (like most editors) doing this in addition to their regular job, and again, the onus on you is to do everything in your power to make the process as easy and painless as possible for them. If you annoy them with little things like sloppiness in figures, grammar or spelling mistakes it will affect their opinion of the rest of your work. Why should they spend their time correcting things that you should have corrected if you cared about your work?

Each reviewer will provide a recommendation to the editor about the paper. The four options they will select from are: *accept* (Woo! Time to celebrate, but very rare for first submissions), *reject with minor modifications*, *reject with major modifications*, or *reject* entirely. The editor will then read all the reviews and decide the outcome. The editor has the power to ignore reviewers; however, most of the time, the editor will consider the substance of the review and use those comments to decide. You will then receive this decision and the reviewers written comments.

The length of time from submission to decision usually takes between two and six weeks, although I have had papers that have taken four months plus at this stage. In the journals I edit for, some papers take a lot longer as it is hard to find appropriate reviewers that will accept the assignment.

8. Revise, rebut and resubmit.

When the reviews are returned to you, your next job is to deal with the comments raised. Many will be minor and just require text edits to clarify things or possibly a written response directly to the reviewers but no change to the manuscript. Others might take months of work in the lab. Also, if your manuscript was sent out for review but then rejected you will still get the reviews and can use those comments to improve your manuscript before submitting to a different journal.

You might be frustrated, upset, and disappointed by the review (I once read reviews on one of my papers whilst at Disney World, I do not recommend this). However, remember that the purpose of peer-review is not to offend you, but rather it is to make the paper better and to make the conclusions more robustly supported. In the long term, you want the public record of your work to be as good as possible and to stand the test of time. Take a deep breath, and read the comments again carefully identifying line-by-line what you need to do to address those comments.

Once you have revised the manuscript, you go through the same process for resubmission as you did in the initial submission. This time around you include a point-by-point *rebuttal letter* detailing what changes you have made. You may also need to include a version of your manuscript where you indicate all the changes.

In your rebuttal letter, be polite but definitive. Be prepared to defend your position while still remembering that you are writing to a real person; upsetting or annoying the reviewer or editor will not help your cause. You might get comments which are due to reviewer misunderstanding aspects of your work. Although it is tempting just to rebut this by saying that they didn't understand, you should acknowledge that any lack of understanding is actually because of the way that you wrote or presented the findings. If the reviewers were confused or missed the point, then so will future readers. If you ever get comments like this, then you should clarify what you meant both in the manuscript and in the rebuttal letter.

Some of the comments or suggested improvements might not be able to be addressed for a variety of reasons. You can write this in your rebuttal letter and then it is on the reviewer and editor to decide if it is justifiable. If a comment is reasonable, then you should always do something about it.

The revised manuscript will go back the same editor that processed the original submission. At this point they will look again and, if the changes were very minor, they may be able to decide to accept it without sending it out for review again. More commonly, the editor will send the revised version, including the rebuttal letter, to the same people. Again, the reviewers might make a fresh set of recommendations based on your improved manuscript, and this time, they will hopefully accept it. However, you may get another round of comments to deal with. The second reviews are often much quicker as usually the same reviewers receive the work and it doesn't take as long for them to process it.

Planning the figure and the paper are good ideas,
planning the Nobel prize acceptance speech is
probably going too far

9. After acceptance.

After you receive the email saying your work has been accepted, and after you have eaten your celebratory cake and drunk the champagne, you are still not quite finished. The journal may now request the original versions of your figures and possibly other details from you like signed ethics forms. They also are likely to request for the original data to be upload to a central repository.

A few weeks later you will receive *proof* copies of the paper. These will look like the final print version, i.e. the formatting will now fit the journal with figures sized and embedded into the pages. The page proofs will contain notes from the copywriter/page setter or editor asking you to clarify specific points, to confirm other changes they have made or to address aspects of style. You should very carefully check the proofs (literally proof read them) because it is more difficult to fix things after they have been published. Usually, the proof reading is on a tight deadline, and you'll likely get only two or three days to get it back to the journal.

Once you send the proofs back, the paper will be published. It may go online immediately as an accepted manuscript, and print copy come later. There may be an embargo period before you are allowed to talk about it and the journal will tell you these important dates.

However, as soon as you are allowed, you should start telling people that to read your fabulous new work. I do recommend investing a little time in self-promotion whenever you have published something. There's no point doing all the hard work of getting the paper out if no one knows about it. There are a variety of ways to promote your paper including press releases if the work will be exciting to the general public. You can also post on your and your departmental social media accounts and to places like ResearchGate and LinkedIn.

6.4   Literature Reviews and Review articles

The first thing you will do when starting a new research project is bring yourself up to speed with the literature. To be able to identify gaps in the current knowledge, you will, of course, need to know what has already been determined. As a new researcher, you will also be reading so that you can design your experiments effectively, understand the data you acquire and will be able to interpret those data within the context of the wider scientific literature. For these reasons, many courses require that you write some form of literature review early in the process to formalize and cement these lessons. If you are writing a dissertation or doctoral thesis, the first main chapter to your work also take the form of a literature review as a comprehensive introduction to your work.

As you progress in your scientific career, literature reviews, in the form of review articles, commentaries or textbook chapters, provide a mechanism to summarize and combine recent findings enabling you to present an updated overview of the current thinking within the field. These review articles provide good entry points for people new to discipline and being invited to contribute a review is a good sign that your opinion is respected and valued by your field.

Differences between a literature review and an introduction to a paper or thesis

Often people consider "literature reviews" and "introductions" as being synonymous, but I encourage you to think of them as separate entities. An introduction sets up the study, whereas the paragraphs within the literature review are the content that your reader has come for and should stand alone in their own right.

It is a slight oversimplification, but an introduction's job is to *map the gap* in the knowledge whereas the literature review's job is to, well,... review the literature. In an introduction, every single piece of information you deliver should be focused sharply on either the question your research answered or to bring in essential information that is required for the interpretation of the data. Anything that is not *directly* relevant to your study should not be in your introduction, and everything in the introduction should be presented with "flags" that help a reader see why it is relevant. In contrast, in a literature review, you will write a much more comprehensive summary of the work; you will delve deeper into the background and go wider with your coverage of the literature. The oversimplification part of the prior statement is that you should also map the gaps in a literature review but, usually, this isn't the primary reason a reader has chosen to engage with your work. Everything in a literature review, of course, still needs to be relevant to the narrative and you should signpost information to help the reader connect the different aspects of the story.

Differences between a literature review and a review article

The type of literature review required for a dissertation or a thesis could become a review article without too much editing or vice versa. However, it is worth pointing out some notable differences that primarily stem from the target audience.

If you are writing to fulfil course requirements, the people reading your work are assessing your coverage of the field, judging whether your understanding is deep enough to warrant a passing grade, also grading your structure and delivery, and looking for ways to help improve your work by providing feedback. Even if your work isn't very good, they *will* keep on reading anyway, it is part of their job (they might not enjoy it though). In contrast, in a review article for publication the end readers can choose whether to read your work *or not*. If those readers don' feel they are receiving some return on the time invested, then they will stop and read somebody else's work.

So, the big question: how can you add **value** to a review article? The good news is that you know the answer to this already. You have read review articles yourself; some you have found to be useful and others less so. The useful reviews deliver some insight that cannot be obtained from somewhere else. Instead of just a stale record of the "facts"*, the good reviews will synthesize and process the combined findings in a way that advances the understanding of the field or in a way that makes the work accessible. In many ways, this point is how review articles differ from textbooks; not only are they usually more up-to-date but also a review should add some valuable insight (there's that word value again!).

Could you read my paper Prof

Beyond the classroom environment your readers are not being paid to read your work, therefore you must establish value through the quality of the content and effectiveness of delivery.

*there are very few "facts" in science. I recommend avoiding using that word.

Systematic Reviews and Meta-analyses

The majority of literature reviews summarizes a topic with a broad scope, taking a qualitative approach where every kind of source can be relevant. They can use biased sources and it is up to the writer which pieces of information are included and how they are weighted in terms of the story. The value of these pieces come from the synthesis and commentary provided by the writer. However, it is by no means perfect. Instead, one could use a formalised approach to literature gathering and perform a quantitative analysis of the data in an attempt to remove the subjectivity of the exercise, increase clarity and provide a statistically supported output. These structured exercises are termed either *systematic reviews* or *meta-analysis* and are a form of secondary data analysis.

Systematic reviews are exactly like they sound. The researcher defines a specific search strategy including the search terms and databases to be used, and strict inclusion and exclusion criteria. They report these defined rules in a methods section, usually with a flow diagram supporting. Almost always you find these in a clinical setting where the objective is to answer a specific clinical question. Usually the results of the search are written in a narrative style but may be supported by numerical data.

A meta-analysis, similarly, takes a structured approach to data gathering but adds a statistical element to the exercise. When the study designs meet the inclusion criteria, the data from the studies are pulled together to generate a much larger data set and by harnessing the increased numbers, produce more robust statistical outcomes. It is possible to do a meta-analysis without a systematic review; however, the outcome will be more robust with defined inclusion criteria. With the explosion of big data studies and the resultant availability of vast databases containing outputs from multiple studies, the scope for meta-analyses is large and growing. Meta analyses are usually written up in the same way as a primary data research paper, with a results section that is focused on the number, and will be more valuable to your CV.

The value of systematic reviews and meta-analysis are that they attempt to make sense of the noise that comes from many studies using slightly different approaches to ask similar but non-identical questions in different sources. The secondary analyses *use that data to come to a conclusion.* However, often the conclusion of a systematic review is that there isn't sufficient quality of data to answer the question and it calls for design of a dedicated study (this is a valuable finding as it can define what questions should be asked). In contrast, the meta-analysis conclusion are supported by statistics and can often go on to inform clinical practice.

Aspects of a good literature review

Three components to a good literature review:

Demonstrates critical insight.
Comprehensive, reduces the workload of the reader.
Enjoyable to read.

Options for how to demonstrate critical insight

Your goal in your literature review is to do more than merely telling the reader what has been done and what is known; you should assimilate that information, combine it and move the understanding forward. Summarizing is easy, insight and synthesis are more difficult at first. If your work is being graded, the *synthesis* portion of the mark scheme is the way you can change an OK grade into a really good one.

For someone starting on a research career, it can be hard to spot where you can add insight. However, the more you read, the more you will become aware of little things that either don't connect fully or where the connections haven't yet been explicitly made. My broad advice is summarize in your first draft (your thinking draft) and while doing so, think about the next three options and what could work.

Option 1, compare the findings and conclusions of older papers to the more recent ones. All papers have limitations to how definitive or wide-ranging their conclusions can be. However, incrementally building up the evidence from multiple papers over many years removes some of those limitations. Experimental systems and techniques have got better over time, and the data from the newer techniques may change the way the older data are now interpreted. This doesn't mean the old papers are "wrong" while the new ones are "right". But it does mean that it *could* be valuable for you to discuss where recent advancements have answered some of the older research questions, or rules out an alternative interpretation. You might also find the opposite to be true; some things were known in the older literature and have now been forgotten or were not considered when interpreting the newer data. Older and newer are, of course, subjective, and this approach can work just as well with papers published close together. Indeed, it can be especially effective when papers come out at roughly the same time without discussing one another.

Option 2, expand on discussions in newer papers. Newly published papers are less likely to have been extensively covered in review articles. There are word limits in papers; therefore, the authors may not have been able to explore the ramifications of their findings fully. In longer-format writing, you have much more space flexibility. Therefore, you might be able to add value and insight by expanding on the points raised in terms of how these new findings fit within the literature. Be aware also, that an author of primary research paper might have taken a slightly skewed view of their work, emphasizing their own prior work and downplaying the prior contribution of others.

Option 3, point out the caveats or limitations to interpretation. Don't do this too aggressively or you will sound very negative. It is better to be selective and only discuss limitations where they actually matter to interpretation. You also aren't really providing insight if you point out something obvious. For example, simply pointing out that something needs validation in a more complex system or in a more generalizable population is not particularly insightful on its own. However, going just a little further by

Big Tip
You can point out limitations without being negative. Emphasize what *can* be interpreted from a paper rather than focusing solely on what cannot.

identifying what the potential benefits of doing these additional parts of research and/or how you could go about addressing the potential limitations are much more useful to your readers.

Option 4, resolve controversy or present your opinion. If your field has some conflict, some differences in opinion, then you can present those opinions and also add your own critical insight by discussing their relative merits and which you consider most likely. You might find this hardest as a new writer, but you can work round any potential fear of annoying your peers by talking about which aspects of each side are particularly strong. You don't have to say that one is 100% right and the other wrong!

What conclusions have you come to about the literature?

I hate all of it!

Even my papers?

Especially those!

That wasn't the critical insight I was hoping for!

You don't have to be negative. Critical insight is about drawing together multiple individual findings to make a more complete whole

Adding value with summary tables

One of the things you must do when writing your review articles is read lots of papers on the subject matter. This will put you in a position where you can provide value to your readers by removing their need to do the same work. In the text of the review your comprehensive knowledge of the field provides value by you selectively extracting the useful information from each paper and combining them to provide a holistic understanding. However, another easy way to add value is to collate discrete points into a summary table. That table can become a useful resource in its own right.

A table works well when you want to deliver general comments in the text but then put supporting specific examples of specific details in the table. For example, if you are writing about a disease, it is likely that you have read lots of case reports. Each case report on its own is not very valuable, likely you won't want to mention specific details of any single patient's data in the text, but when presented together, they might combine to tell a more interesting story. A table of clinical features, genotype to phenotype correlation or outcomes in relation to treatment could all be useful (to a reader. Always include a column with references to the primary literature as part of these tables.

Make it enjoyable to read: tell a story

The first draft of your literature review is likely to feel like a list. If you leave it like that it will be brutally painful to read. Lists are boring. Your grades will be much better if there is a narrative feel to the information you are delivering. Top marks if your marker can read your lit review without falling asleep.

Stage 1: identify connected findings. One common problem is that new writers often discuss individual papers in isolation. Don't do this. Make notes on the papers you are reading, identify what they have done that is going to be relevant to your story and then try to look holistically into how the studies connect to deliver the point.

Stage 2: assemble your connected studies into a narrative. There are many ways to do this and what you decide should be based on what you want the reader to take away from a specific subsection.

Simple yet effective options are to present paragraphs or subsections as either a journey through time or as a point/counterpoint argument. The second method is particularly effective when you are building toward the research questions your thesis addressed, or where you will go on to discuss future directions to address this conflict.

Examples of narrative styles

The journey through time.

The importance of (*something*) was originally revealed through the study of (*e.g. a disease, a population*). Specifically, (*mutations, protein, people*) were shown to (*action, phenotype etc.*) through (*analysis technique*). Around the same time, different groups were investigating (*something related*) using (*a slightly different model system*) this provided additional insight into (*part of the study*) but raised questions as to (*limitations*). More recently, (*a different approach*) added (*further detail*)…

The Point/Counterpoint.

There are two major schools of thought when it comes to (*something specific*). … is believed to be … by … This is supported by (*approach*) which demonstrated that (*experimental specifics*) there was (*experimental outcomes*) and by (*different approach/series of studies*) which addressed the question by (*experimental description*).

However, an alternative interpretation has been presented where (*interpretation 2*). This was based on (*experimental system and data*).
Real Example:
Although the pattern of laminin-matrix deposition in diverse cultured cells is quite varied, it remains to be determined whether laminins in intact basement membranes in different tissues exhibit distinct patterns of deposition. Morphological studies of basement membranes and studies of laminin-subunit expression in tissues generally involve conventional electron microscopy and light microscopical immunolocalization of laminin antibodies in sectioned material, respectively. These techniques provide a `yes-or-no' answer as to the overall gross integrity of the basement membrane, including whether it has undergone duplication or focal loss, and provide a readout of its laminin-subunit composition. However, they do not permit high-resolution evaluation of the specific patterns in which laminin heterotrimers are incorporated into the matrix. This might explain why there are no reports of defects in laminin-332 deposition in the basement membrane of the skin of individuals whose keratinocytes lack β4 integrin, despite evidence that laminin-332 is deposited aberrantly in the matrix of keratinocytes derived from such patients in vitro (Sehgal et al., 2006; Vidal et al., 1995). Of course, aberrant deposition might be a secondary consequence of altered migratory properties of the cells and might not be manifest in tissues in vivo. Immunoelectron microscopical analyses of laminin localization in tissue and/or the development of new high-resolution methods are required to resolve these apparent conflicts.
From: Hamill et al., Laminin deposition in the extracellular matrix: a complex picture emerges, Journal of Cell Science, 2009 122: 4409-4417.

Figures and diagrams

Literature reviews can be very dry, so whenever someone asks me if they should include a diagram, I almost always say yes. Visual representations of textual information are good ways to reinforce points and explain complex findings.

Be aware that you cannot reproduce primary data or any other figure from published work in your literature review unless you have copyright agreement to do so. This applies to your own published works too. Instead of reusing a figure, you should look to assemble new diagrams or cartoons that will help make complex parts of your story easier to digest.

With that in mind, there are a couple

Big Tip

The figures or diagram are used to add to the text.

You shouldn't use a figure *instead* of describing something.

of quick considerations:

- **Only include a figure when it adds value.** Sometimes text is all you need. Do not make a figure just for the sake of it. You likely do not need a figure for really common knowledge or simple things.

- **Integrate the figure into the narrative.** A diagram should illustrate what you have written rather than replace the text. You should provide the set up/connection needed to bring the diagram into the storyline and should not make a new point in a diagram that is not described in the words. You must always reference the figure in the text in the same way you would for a research paper (Figure 1).

- **Create your own (if you can).** Assuming you have the time to generate the figure you want, and the ability to make it look good enough to use, it is always better to create your own. There are software and apps that can help you such as BioRender. Making it yourself means that the diagram is *exactly* what you want, contains the material you need and nothing more (see the next point). In assessed work, doing it yourself also says that you care about your work and shows to your examiner that you understand what you are presenting. If you do decide to re-use (check copyright!) or modify someone else's figure, make sure you cite the original source and are describing every aspect of the figure.

- **Simplify.** Figures are supposed to make things easier to understand! Avoid making or using something that is more complex than you need. If any details aren't relevant to the story you are telling, then simplify the diagram so that they don't dilute the point you are trying to convey. Remember that you should be describing everything in the figure.

- **Use unpublished data from previous studies.** Your narrative might benefit from including representative images (often microscopy images). In these situations, if your lab has published in a related area, then they almost certainly will have extra images that weren't used in publications. These images might not be tied up in copyright laws and may be available for you to use. Think about what will make your work stand out, be more engaging and help to tell the story, then ask if appropriate pics exist.

6.5 Referencing

You need to provide the references to **all** material used in preparing your literature review or research paper including any figures. Failure to cite effectively will mean you assignment gets flagged for plagiarism and runs the risk of failing the assignment or, worse, being expelled from your University. For work submitted to be published, your manuscript will get screened and rejected without review.

In practical terms, the standard approach in science writing is to indicate where our information came from directly in the sentence by providing in-text citations (1, [1] or Author et al., depending on style), then provide a complete reference list at the end of the main text of the manuscript. Using footnotes as a means to cite is less common.

What should you cite?

You should cite everything that is not either definitively yours or which is not classed as *established knowledge*. If you are even in doubt, cite. Better to have too many citations than too few.

There are really three types of citable material:

- **Summaries.** Many of the sentences in your paper will be of the "summary" type. You will have read one or more primary data papers, pulled out the key findings relevant to your work, and then written them in your own words. Although you have assimilated the information, the sentences still require citations every time. You must indicate where did the original information or idea came from. Failing to cite is essentially saying that you did the work. The citation provides a route for the reader to make their own decisions and interpretations about the source material. They may interpret those published data in a different way from yourself.

- **Paraphrases.** To paraphrase is to restate the same fundamental point as presented by others but in your own words and own sentence structure. You must cite the source every time you paraphrase; rewording something doesn't make it yours!

- **Quotations.** You don't see many, if any, quotations in scientific papers. It is much more common to paraphrase or summarize. If you do choose to use identical phrasing from published work, the phrase needs to be surrounded by inverted commas and the precise citation provided, usually including the exact page number where the quote has come from. Quotes are rarely the best way to convey your message (by definition, it is not your message if you use someone else's words! The only times I consider a quote are where it is important that the reader knows that I am using some other author's precise wording i.e. to deliberately identify a short and specific phrase to someone, usually it is a phrase that I disagree with and will go on to rebut.

How do you cite?

In text citations

Most in-text citations should be placed in parenthesis at the end of the sentence where the citable material is located. Those brackets are put either just before the full-stop if it relates to the whole sentence or immediately after a sub-clause if the reference only supports that part of the sentence. "Piece of information 1 (Author et al., year), piece of information 2 (Author et al., year)".

There is a formatting exception if you choose to make the author's names part of a sentence. In these cases, you provide the year indicator only; "In work by Author and Author (year), …". Be aware that sentences constructed in this way place emphasis on the author of the source, which is often not what you need to emphasise. Usually the sentence should be focused on the point being delivered. Therefore, I recommend using "Author et al.," sentences sparingly and only in situations when the author name is actually relevant to the narrative.

Big Tip

It is easier to see that you have cited the correct source material if you keep your citations as (Author, year). Save the final formatting until the end.

Quotes or very specific pieces of information from within a large source may require a limore information in the in-text citation, and you may need the page/line numbers; "…quote…" (Author et al., year, pp XX-XX).

There are many ways to format the in-text citations (and the reference list). For example, how many authors are listed before saying *et al.*, or whether to use italics, superscript numbers etc. Which method to use depends on the style guide from the journal or will be defined by the assignment so, as usual, check your instructions. If your assignment doesn't tell you what to do, then use one of the standard formatting systems (Harvard-style referencing is common for science writing).

Example in-text citations

Nature

Single-cell RNA sequencing can reveal RNA abundance with high quantitative accuracy, sensitivity and throughput[1].

Cell

Wild-type (WT) or four mutated derivatives of pri-miR168a (m1-4) were expressed from the Cauliflower mosaic virus 35S promoter in the Arabidopsis mir168a-2 mutant (Vaucheret, 2009).

Reference list

In addition to the in-text citations, you will require a reference list. In most manuscripts, this list appears after the discussion and acknowledgements but before the figure legends.

Again, each journal has its own style for how references should be formatted. The good news is that there are reference manager programs available (EndNote and

Mendeley are two of the big ones) which store all your papers but make formatting easy. Hopefully, you are using one of these programs already. If not, now is the time to start. These programs come with many journal styles already preloaded but if they don't, you usually can download the style requested by the journal. Once you have the style, you need only apply it your manuscript, and it will format all the in-text citations and generate a reference list with a single button click. They save you a lot of time compared with doing it by hand, especially if you change journal and therefore need to reformat.

If you aren't using a reference manager then the things to look for are the order (alphabetical or in the order they were cited), whether the references are numbered or not, the number of authors before et al., which aspects of the title, journal name etc. go in bold or italics and the order the journal, title, page numbers etc. are presented.

Even if you do use a reference manager, do a manual check of the list after you have inserted it as sometimes errors do occur. Errors are infrequent but usually they occur when something has been inserted into a library twice (check your ref library for duplicates) or where the record in the reference library has been corrupted in some way. Be especially careful with formatting, e.g. Greek symbols, might get changed if you perform a document-wide conversion.

Example Reference list entries

Nature
Linnarsson, S. & Teichmann, S. A. Single-cell genomics: coming of age. *Genome Biol.* 17, 97 (2016).
Cell
Vaucheret, H., Mallory, A.C., and Bartel, D.P. (2006). AGO1 homeostasis entails coexpression of MIR168 and AGO1 and preferential stabilization of miR168 by AGO1. Mol. Cell 22 :129–136

Cite while you write or close the books and write?

What I find works best is to do lots reading *before* I start writing. Then, once I have a good idea of what I want to say, I close my books (or, more accurately, minimize the tabs with journal articles in them), and then I sit down to write a first draft without looking at them. Remember that the first draft is all about getting the core framework of the story in place. You want to write rapidly and as fluidly as possible rather than stopping every sentence to find and insert the citation.

Once I have the core framework in place, I start

Incomplete referencing can have serious consequences. Not usually corporal punishment, but it is a form of plagiarism and could seriously harm your grades!

editing and adding the more specific details; those details include the references. When writing my first draft, I sometimes include a note to remind me to insert references later, usually with a coloured highlight (ref) so that I can quickly spot areas that need attention. The more established you get in the field and the better acquainted you become with the literature, you will begin to remember the author and year of the paper that you intend to reference. In these cases, you can directly cite from your library or add a more complete note (Author 2022).

If you do wait until the very end to insert references, make sure you leave plenty of time before your deadline. It always takes longer than you think. While reading, ensure to take good notes, you will need to know where each citable idea has come from; these notes will make the citation process less arduous.

I want to take this opportunity to preface a really point: titles and abstracts are extremely important! During the referencing process, you will likely need to search through your reference library file or connected online database looking for something to support your point whenever you cannot remember exactly where the idea came from. If you can picture yourself doing this, then it is no stretch to imagine other authors doing the same thing. When it comes to choosing a title and writing the abstract for your manuscript you want to be sure that *your* paper will appear in other people's searches and that they can identify the message within your paper correctly. You do this through a combination of having a title that captures the conclusions of your paper and an abstract that contains keywords that people will use in their search when looking for related material. We will return to this point in the next chapter.

Referencing: what is expected in each sub section

Abstract: usually, no references

The only time you include a citation is if your work stemmed from a single prior study, or when two closely related papers are being published alongside each other in a special issue. Citations rules in abstracts are different from the main body of a paper; you use the whole reference as your in-text citation (authors, title, journal, and page numbers). Including a reference in an abstract eats up a lot of your word count, so my really strong recommendation to try to avoid situations where you need to include one.

Introduction: lots of references expected

Almost every sentence of your introduction will need to be supported by at least one citation, the only exceptions are possibly the paragraph topic and wrap sentences (but they might need something too). Book chapters or, more commonly, review articles might be enough for the first paragraph. Once you get into the more specific details of the introduction, I would expect to see primary data journal articles throughout.

Literature Review: multiple references per paragraph

No surprise, you need to cite the literature! Every paragraph will need to contain multiple references. Indeed, many individual sentences will need require more than one citation. As an examiner, if I see a paragraph without any references, I automatically start wondering why not; is it plagiarism or is it really entirely the student's own thoughts.

Book chapters or review articles can be appropriate sources for core concepts near the beginning, but you should not reference a review to make a specific point. I also recommend flagging these by using (as reviewed in...) in the in-text citation. Use primary data anywhere where there are specific details. Especially in dissertations and theses, where are not limited in terms of numbers of references, you have no reason not to cite the original work.

Methods: sources and validation

You may need to include references for where protocols and reagents came from or where reagents were tested and validated. For example, you might have used things like cell lines, antibodies or PCR primers that were described elsewhere, or it could be that the original designs for survey questionnaires and analysis methods may have been previously reported. Mostly you will use primary data articles here; however, some excellent methods books are also widely used and cited.

Results: occasional citations to help transitions

You will need references for any transition sentences where you are establishing the rationale behind an evolving hypothesis (more on this later). These points are likely to be very specific, so I would not expect any reviews, it will be primary references only.

Discussion: lots of references

You will need many references throughout this section. Every paragraph will have several references. Indeed, if you have a paragraph that doesn't use any outside sources, then it may be an indication that this part of your discussion is not effectively connecting your data to the rest of the published literature. Most of the time, you will be referencing primary data publications as the core concepts will have been established in the introduction.

Figure legends: adapted from?

No references needed for your own work. However, if you have adapted a diagram from previously published work then you must provide the relevant citation(s). *Adapted* is an important word in that last sentence. If you directly reproduce published artwork you should ask yourself two questions; i) do you actually need the figure, or would the citation be sufficient? and ii) do you hold the copyright or have the appropriate permissions to use the image?

What to check when editing

No surprises here; make sure that everything that needs a citation is cited!

The most commonly missed citations are those where there should be more than one citation for the material within a sentence. This happens as, when you are editing the paper, those sentences look like they are complete as at least one citation is present.

Specifically keep and eye out for are sentences where you have said "studies", "widely supported", "well established" or equivalent phrasing indicating that the idea has originated from more than one source. In these cases, you shouldn't be citing reviews unless you specify "as reviewed in". I recommend adding a note (more refs) in your draft to make sure you don't forget.

The other common problem is where the same source material is being used across multiple sentences. This is fine, it happens a lot, but make sure it is clear from the phrasing that the material is linked. If it is not clear, it is safer to re-cite than risk appearing to have failed to attribute the source.

Never carry linked citations across paragraphs, always re-cite in the next paragraph. Any paragraph without a citation is very likely to be noticed.

Plagiarism checkers

Universities and journal will run all submitted assignments through software that checks for plagiarism. This should not be a concern for you if you have cited all your sources. The software will generate a report and your markers will be able to see the similarity between your work and every other piece of work submitted to the system and to everything available online. They will be able to request the original works automatically via the system and compare side-by-side. Being similar doesn't mean it is plagiarised, therefore the person looking at your work will make a judgement call. You can check your own work before submission, but if you have done things appropriately, there should be no need.

Et al. is an abbreviation of the Latin, "et alia" meaning "and others". Your target journal will define whether Latin terms like this and in vitro, ex vivo etc. should be italicised or not

6.6 Writing for the Public

Communicating science to the public is becoming a much important part of a scientist's work. One of the goals of science communication is to break down the barriers between research within the "ivory tower" of academia and the general public. We are living in an era where scepticism of experts has been encouraged as a means to promote political agendas and "where controlling the narrative" of news stories is more effective in shaping decisions than critical appraisal of the data.

Funding for non-commercial research comes from the public either via charity donations or from government spending of tax income. A shift in the political winds has knock-on effects to funding availability and to scientific progress. It is, therefore, more important than ever that we, as a scientific community, publicise our work widely and effectively to promote the value that investment in research brings. Indeed, most grant proposal require lay summaries exactly for this reason; to show the public or donors where their money went, and these are often used in grant assessment.

As a field, we are also recognising the importance of public discourse in terms of the outputs from our research. Many journals now require you to include some form of summary that is suitable for a non-specialist audience (a *lay abstract*). These are already quite common in the bigger, cross-discipline journals and are gaining in popularity.

There are lots of other mechanisms to get your story out there. Writing press releases, contributing to news articles, and writing blog posts are all ways that you can easily increase the reach and impact of your work.

Why bother?

- **Enjoyment.** Most scientists who write or speak to the press and public actually enjoy it.

- **Funding.** More media or more coverage means a higher profile for you, for the department, for the University, all of which can help lead to future funding. Communicating to the public is not purely an altruistic endeavour.

- **Citations.** More media coverage means more reads and that can lead to more citations.

- **Ethics.** Your funders should know how you have spent their money, especially if the money is coming from the public purse. The people who gave the money should know that where it has gone.

- **Advocacy.** You research could help to inform government officials, funders etc. as well as the general public. You can change society, improve people's lives. The importance of scientist to politician discourse was widely apparent during the pandemic. Being able to communicate in a way that the policy makers can understand is critical for the most appropriate course of actions to be selected.

> - **Practice.** The more frequently you write anything, even when it is for a different audience, the better you will become at writing in general, and the easier you will find getting future papers and grants written up and accepted or funded. Writing for the public forces you back away from the minutaie and focus the *story*. You can bring these lessons of identifying and emphasising value to your scientific writing.

Who are you writing for?

As always, you must write for your audience. Identifying your audience before you begin is important for the same reasons as identifying your academic audience, you need to know what the readership will value within your work. Before you write a single word, you need to be clear on your objective.

Lay abstracts and news articles are usually aimed at an *intelligent 14-year-old* (according to the BBC), someone who has some basic science teaching but who will not have any specialist knowledge of the field. This means that you will need to deliver most of the concepts in more accessible terms and you should not assume too much prior knowledge. This does **not** mean that you should completely dumb it down. For examples of the style you are looking for; have a look at the science pages on BBC News or NPR.

If you are working on a human or animal disease or an environmental concern, then it is likely that your audience will know quite a lot about specific aspects of the situation but may

If your work is going to make a difference in the real world, you will need to communicate with the public. Thankfully, most people are actually very receptive and interested.

know very little in other areas. Patients with a condition will very frequently have engaged with online resources such as Wikipedia or WebMD before encountering your work. If you are asked to write for a charity or one of your funding bodies, then it is worth having a look at their promotional material and speaking with people in that charity to work out how to pitch your story.

Remember that scientific literacy changes over time. Again, the pandemic provides a good example of this. In pre-pandemic times, I would have felt it necessary to explain key concepts about PCR, antigens or viral packaging, whereas two years later I would start from a position of assuming more specific prior knowledge.

Whatever your write, always get a non-expert within the target audience to read your work. I, genuinely, still ask my mother to read my lay summaries before I submit them!

Word Choice

Unsurprisingly you will need to change your language, sentence and paragraph structure when writing for a lay audience. No surprises here, you should use the simplest, most accessible language possible throughout. However, if you set things up properly and build in complexity at an appropriate pace, then you can deliver complex concepts without any issue at all. Just take it slow and surround new ideas with things that your readers will already appreciate..

It's hard to know what phrases are suitable for a lay audience so get input from someone who'll give you honest input

Structurally, the best approach is to use shorter sentence with one or two clauses only and a maximum of 15 words per sentence (for most sentences). Avoid using abbreviations or acronyms, use the whole name throughout, unless the acronym is already used by the target audience. Avoid complex or meaningless terms and phrases, e.g. 'virtually', and don't use a long or the fancier sounding word when the shorter more commone one will do just as well e.g. choose *use* rather than *utilise*. Don't use *significant* as a stats term, talk about confidence and don't use it as a term for importance or magnitude term, it is better to talk in specifics. Note that most of these points are good general advice for writing and are not specific to writing for the public!

Use person-centred language throughout. Focus on the person, not their illness 'people with a disability' is preferable to 'the disabled'. A person 'has glaucoma' rather than 'is a victim of glaucoma'

Content advice

The human angle is always good when you have that option. In quantitative science, we get hung up on population statistics; however, using a specific individual example can be an effective strategy for helping your message resonate with the audience. You will see examples of this with every documentary or news report, the infographic or population-level information is always supported by a personal interest story to provide the hook that tugs on the emotions of the audience.

Big Tip

Use person-centred language in your writing.
A patient is not defined by their illness.

For complex or more abstract material, consider using metaphors to make it more accessible. There is a danger of being cliched here, don't overstretch your metaphor, but providing a reference point that makes sense to your readers will allow more abstract concepts to be more easily absorbed. On this point, I find it tempting to cut back on the description of all the caveats around my work to make it simpler and more direct. However, you really must be very cautious in this approach; this sort of simplifying could easily make your message stronger than your evidence can support.

Say thank you. Always remember to credit the funders, the patients or participants. Being aware and acknowledging that research can only happen because of the support of the people is important.

The next few sections talk about what to think about in specific types of writing.

Lay summaries (for grant applications)

Grant applications usually ask for a brief summary of a research project that has been written for members of the public rather than researchers or professionals. Lay summaries can be really important in getting a grant funded. They can be required as part of a grant's conditions or when recruiting for trial participants, and are often used on the funder's websites to show their donors where the money goes. For some charities, grant applications must first be passed by a panel of lay members who comment on and score each application on its importance and research priorities, before moving on to scientific peer review. Patients or carers groups are also increasingly involved in helping to create the funding applications in the first place. They have a direct interest in the outcome of the research and experience of the disease and can advise on the feasibility of the research. The lay summary helps these invested individuals understand the research and enable them to take part in the decision-making process.

Specific requirements differ from funder to funder, so make sure you match their guidelines with respect to structure, information required and word count.

Things to include in lay summaries

Motivation

Why the research is being proposed.

Aims and objective

What you will achieve.

Impact

How achieving the objectives will improve the situation for the target audience, the research community and the rest of the world. Also, make the timescales clear

Risks

Make sure that any risks attached to the activity are clearly identified and described. If a reader identifies a risk that you haven't talked about, you will seem naïve.

Dissemination

How are you going to get your results out there? How will you engage the patients, politicians or other relevant invested parties?

The list above sounds like a standard grant summary. However, although the concepts are very similar the angle you should take is different. The challenge is making the information easy to read. Use short, clear sentences broken up into paragraphs for readability, and avoid complex grammatical structures. Don't be demanding of the reader and don't ask them to remember too much jargon or abstract information.

Put the scope of the grant front and centre. For example, you might start by saying how many people have the disease or what it cost to treat, or what the societal implications are and how you will contribute to dealing with these issues. For your aims and objectives, put extra effort to be explicit in why completion of these aims will be valuable. Use concrete, everyday examples wherever possible. You do need to describe what you will actually be doing in a way your readers will understand. This includes your study design, methodology, expected outcomes and overall strategy. But you do not need to go too deep, an high level overview is usually better unless you are specifically developing a new approach. Instead of putting emphasis on the experimental procedures, instead put maximum focus on the motivation and impact parts of the summary.

Lay abstracts (for papers)

A lay abstract for a paper is similar the scientific abstract (discussed in next chapter). It consists of motivation, background, aims, methods and results, conclusions and implications. Usually without subheadings. However, for a lay abstract, I recommend using a greater proportion of the word count on what the study means; the motivation, conclusions and implications parts, whereas for scientific abstract, you will need to focus more on the middle sections, the methods and results.

A recommended breakdown for the lay abstract is to use around one-third of your word count in setting up the study, about one third covering results with very minimal/superficial coverage of the methods, and the final third devoted to what impact the research will have in the wider world.

Big Tip

Put the greatest emphasis on the motivation and impact aspects of your study in your lay abstract.

Press Releases

Not every piece of work you do will garner press attention, it's called 'news' for a reason, but your work is even less likely to be noticed if you don't tell people about it! If your story is new, quirky or features amazing results, or when it contains simple results that are impactful, then it can be valuable to prepare a press release. You will have a better chance of your story being picked up if it has some human-interest element, is relevant to the current news agenda and has strong visuals to help sell the story.

Journalists want to entertain, inform and educate *in that order*. As always, pitch your story or write your piece to reflect the news outlet, e.g. local radio is different to the science pages of a major newspaper and as such you would need you to take a different angle. Reducing the work required for someone to run a story increases the chance that they will choose to use it. If you can, write it like a reporter for that outlet would. This will make it more likely that the story that needs minor edits rather than one that needs to be completely re-written. Your institution will have a press office and usually, your first step is to approach them to help you get the work out. They will help you to identify how to make your work newsworthy, and who is likely to pick up the story. You will probably have to provide the first draft, but they will help tidy up and improve the targeting of your message.

In terms of structure, there is a simple message: front-load everything. Unlike science writing, where the big conclusion comes at the end, press releases are structured so that the first paragraph grabs the attention by summarising the whole story. After the summary paragraph, you can build toward the finer detail. When it comes to press releases, it is only the committed readers that will keep going to the end.

With press releases, it is important to be prepared. The good news is that the processing timeframe of a journal article means that you will have a good idea when your work will be coming out. Some journals establish an embargo on press releases, they will tell you about this as part of their correspondence, this will give you time to get everything in place on schedule. When it comes time to send out your press release, do not send it to the general email address unless there's nothing else. Instead, find out who the most relevant health/science/ correspondent is. If the release is about a published study, send the original paper or a link as well. If you have images or videos, you can use send them as well so it can go straight to print.

Blog posts

Blogs can be an effective way to build up an audience for your work. Having your own blog or contributing to others can be a good way to get into the habit of writing regularly. The more you practice, the better you become. The easiest can be like a diary or journal of your lab, but if instead, you want to build a wider readership, then you will want to identify content that people will find useful. That useful content could be entertainment, advice, educational or reviews.

Blogs are meant to be conversational, so use a friendly, informal tone, although how formal varies with the blog and the audience. If you are contributing to your University departmental blog or to a charity web page, you are likely to be more formal than for a personal blog.

Structure and content

Use an inverted pyramid of information. Have an intro/summary paragraph with big ideas at the top, then build toward increasingly fine detail as you work down (same as for a press release). You may also have a preview paragraph that will appear on search engine results, so your introduction becomes doubly essential if you want people to click on your story.

Usually, the content on a web page should have much shorter paragraphs than you would do in prose writing. Imagine your page being read on a bus or train journey, large blocks of text can be off-putting. It is usual to see one or two sentence paragraphs, or at least line breaks every couple of sentences. Add structure to your pages; search engines like subheadings, bullet points etc. and structure helps with readability. Use images to break up bodies of text.

After you have prepared the content, check that it will work on different sizes of device (mobile/cell phone screens, tablet and desktop).

To help your page get found and read, you should try to optimise your content so that search engines will place it higher. The algorithms change regularly so have a quick search for *search engine optimisation* to get the most up-to-date guidance. However, the general advice is to identify a phrase or keywords that your target audience will use in their search terms, and then ensure that you use that phrase in your, title, image alt tags, meta description and in the body of text. Don't overdo it; *keyword stuffing* can lead to your page not being indexed. Other things that can help with search engine optimisation is to use out-going links, links to other webpages, or including videos and images.

What to check when editing

Is it lay enough?

No surprises here, the biggest challenge is getting the pitch right, so the best plan is to get your abstract read by someone else. The other common problem is that writers sometimes forget to emphasise or make explicit the importance of their findings in the real world, don't forget this bit!

Chapter 7: Writing Papers

7.1 Step by step to paper writing
Overview of the stages from finishing lab work and data analysis to publishing in a scientific journal.

7.2 The Title Page
All the information required on page 1 of your manuscript.

7.3 Choosing an effective title
The most important part of your paper? Probably. Improve the chance of your paper being found, read and cited by giving it an effective title.

7.4 Scientific abstracts
How to draw your reader in with impactful abstracts.

7.5 Introductions
Set up your study with a focused introduction.

7.5 Materials and Methods
Deliver the details of how you performed the experiments so that others can repeat them, and so your readers can interpret your results.

7.6 Results
How to report your findings so that they are clear to the reader and enjoyable to read.

7.7 Discussions
Putting your findings together to make sense of the world.

7.1 Step by step paper writing

This chapter is broken up into each of the sub-sections of a manuscript. My advice is to read the pages relating to whichever specific sub-section you plan work on immediately before making your first attempt at drafting that part of your manuscript. Then have another quick read through that section again, before you perform your first major edit.

Research papers are the main method of communicating science, this sort of paper that is the most valuable to your CV. Unsurprisingly most of your efforts will be directed toward getting the next paper published and most University0 courses will ask you to write your work in this style as an important part of your training.

There is a skill to writing your paper in a way that the field will respect the work while at the same being appreciated by the readers as easy to absorb. If you can deliver your findings effectively, your work will have a much better chance of attracting citations and thereby influencing the thinking of scientists in your field. As with all things, you will become better at preparing manuscripts the more your practice, but, at first, it will take quite a while to get a draft ready for submission..

The order in which you prepare the different elements of the work is up to you. However, a linear path from abstract to conclusions is unlikely to be the most efficient or effective. For example, what information is needed in the introduction depends on what you are actually introducing, so I find it is usually better if this is one of the last parts you write.

On the next couple of pages is an example of the order in which I write my manuscripts. I've highlighted which chapter to refer to at each stage, including the ones you might have read already.

Many sentences in your introduction and discussion will need to be supported by multiple references. While writing your first draft, close the books and papers and focus on getting the structure correct.

Preliminaries

1. Read background material, design experiments.
2. Collect and analyse data.
3. Read syllabus/instructions, read *General tips (6.1)*.
   - o Set up a document template.
4. Read *Materials and Methods (7.6)*.
   - o Draft materials and methods section.

You can do these steps while you are working in the lab!

First Draft

5. Read *Titles (7.3)* section.
   - o Write a draft title.
6. Read *Scientific Abstracts (7.4)*.
   - o Draft abstract (very rough; focus on motivation, aim, conclusions).
7. Read *Figure Preparation (5.1)*.
   - o Make figure panels from each set of data.
8. Decide on which figure panels are needed to tell your story.
   - o Decide on panel order and which figures should be main text and which should be in the supplemental figures.
   - o Assemble complete figures.
   - o Get feedback on your figures.

Decide the story you want to tell

9. Read *Figure Legends (5.4)*.
   - o Draft figure legends.
10. Read *Results (7.7)*.
    - o Draft results section.
11. Read *Introductions (7.5)* and *Discussions (7.8)*.
    - o Outline introduction: write topic sentences.
    - o Outline discussion: write topic sentences.

First drafts don't have to be word perfect!

12. Read any extra background material.
    - o Complete the draft of the introduction and discussion.

Focus on getting the structure right so that the story flows

13. Update abstract (add methods and results detail).
14. Get feedback on title and abstract (from co-authors).
    - o Update title and abstract
15. Read *Lay Abstract* (in *Writing for the Public 6.6* if required).
    - o Draft lay abstract.

Editing

16. Read *Editing (Chapter 8)*.
17. Edit round 1: does your story flow logically?
    - o   Restructure or reorder if necessary.
18. Edit round 2: check each paragraph.
    - o   Do topic sentences tell the story? Have you included a wrap? Check paragraph lengths.
19. Edit round 3: focus on minutiae.
    - o   Check refs, labels and units, abbreviations, methods, and other consistency checks.
20. Edit round 4: focus on sentence structure and word choice.
21. Get more feedback on title and abstract (from co-authors).
22. Get feedback on the rest of the document.
23. Get feedback on title, abstract and figures from people not familiar with your work.
24. Conduct more rounds of editing / feedback until everyone is happy.
25. Final Checks: read the manuscript backwards to help spot typos and small mistakes.
    - o   Check references have formatted appropriately.
    - o   Perform a final double check and edit of the abstract.
26. Confirm co-authors are happy with final version.*
27. Upload to journal manuscript.
    - o   Wait and hope for favourable reviews.

You must do this before you submit.

Science is a team sport

In the list of steps, you can see that I've indicated lots of points for asking for feedback. If you are writing your very first manuscript, then I recommend getting even more feedback than this, especially early in the process. Don't wait until you have made 7 figures, written 7 figure legends and written all the results before finding out that your supervisors don't like your writing style! It is better to write one complete subsection, get feedback on style and then adopt the lessons learned on all the following sections.

Your co-authors should help you to construct the manuscript and be involved throughout. Indeed, it is likely that other people will contribute individual experiments to the manuscript and you will likely need them to write up the methods and results relating to those data.

7.2 The Title Page

When submitting work for publication, you need a title page containing important details about the paper They are relatively self-explanatory, but a few points are worth being aware of. Note that the exact layout, location within the document of these sections and what is required within each one is journal specific, check the instructions. All these parts take very little time to prepare, and you can do them at any stage of the process. I find them useful to break the inertia barrier and "get started".

Common Features of a Title page

- **Title**: no surprise: see section 7.3 for how to choose an effective title. *Check for character or word limits.*

- **Running title or short title**: a short form of the title. This will appear as a header at the top of each page of the final published article. Choose the main conclusion or the focus of your story. For the running title, I sometimes include a different keyword or synonym than in my main title, to help ensure the paper gets picked up by different search terms. *Check journal instructions for character or word limits.*

- **Author list**: *check the journal requirements for whether you should use author initials or full names and whether you need to include qualifications after names.* Always check with your co-authors what form of their name they prefer to use and ask which middle initials they use.

> **Big Tip**
> Check the name **and initials** your co-authors use in publications.

Who should and who should not be on the author list is a little bit complicated, so I have described that in more detail below. Author order is also important and should be discussed with your co-authors. There are some international, culture and discipline-specific differences in interpretation of relative position on the list.

In biological and medical disciplines within the US and UK, the first author is usually considered to be the person who has done most of the actual data acquisition and analysis work, whereas the last author is usually the person in charge of the lab, had the main idea for the project, obtained the funding and/or supervised the project.

- o **Co-first, co-last authors**: it is increasingly common, especially in larger papers, for more than one person to have made substantial contributions to the final version. In your author list, you can indicate equivalent contributions by using a superscript character (usually a *) after each of the names and include a phrase such as "these authors contributed equally to this work". If you have a shared author paper, you should use the same system to highlight it on your CV (do not change the author order).

- o **Affiliations**: use superscript characters (*check the journal* *instructions for whether numbers or special characters*) to indicate where each author works. Immediately below the author list, write this list of institutions using the numbering scheme you have defined.

- o **Corresponding author**: identify the person who people should contact about the manuscript. Often this will be the first or last author, but the corresponding author is another position that carries a little bit of extra credit, and it may be appropriate to share the credit amongst the team. Again, discuss this with your supervisor and co-authors. Ideally, the corresponding author should be someone who does not intend to move institution soon as they are the ones who will receive any contact about the paper post-publication.

- **Corresponding author contact details**: physical address, email, and phone number of the person designated as the corresponding author.

- **Keywords**: provide a list of (usually five) keywords or short phrases that indicate the main topics of your work. These keywords help search algorithms locate your paper, so think about the people who you want to read your work and what they will type into Scopus, PubMed or WebofScience. Think about what you would use to search for papers related to your work and pick those. Most will be obvious, but don't forget to think about alternative terms for the same thing. You should try to include the same keywords in your abstract, title and, even better, the subheadings of your manuscript; these all helps to maximise visibility to search engines and will push your paper to earlier pages on Google Scholar or equivalent.

- **List of abbreviations**: a list defining all the acronyms and abbreviations used in the manuscript.

- **Funding:** details about who financially supported the work might need to go on the author page, or it might be within the acknowledgements section, check the instructions. This information absolutely must go somewhere in the manuscript.

- **Conflicts of interest**: include a statement declaring any real or perceived conflict of interest, or absence thereof. For example, if you were funded by a commercial entity, you must declare this connection. Be definitive and explicit of what was funded and what that means in terms of limits to interpretation.

- **Other things:** you might need a dedicated ethics statement and/or a data availability statement. You might also need a list of key findings to go in the journal table of contents. As with all other things, check the instructions!

Who is an Author?

What contribution is required to warrant inclusion as a co-author and the order the different names appear in the list is an area where everyone will have an opinion and is probably the biggest cause of long-term fallouts within the scientific community. This is not at all surprising; publication outputs are a key determinant of career progression. Usually, it is best to have authorship discussions early and then again frequently throughout the process. Moreover, be aware that author orders might need to change as the article proceeds through peer-review and additional work performed meaning the relative contribution of different people change.

It is very important to tread carefully, to be open and transparent with decisions, and to follow a strong ethical code when deciding who is an author. Most fields have established a consensus framework for who should and who shouldn't be credited. In most cases, these broadly match the benchmarks established by the International Committee of Medical Journal Editors (below).

To be credited as an author, you must:

Make substantial contributions to the conception or design of the work; or the acquisition, analysis, or interpretation of data for the work.

AND

Be involved in drafting the work or revising it critically for important intellectual content.

AND

Grant final approval of the version to be published.

AND

Agree to be accountable for all aspects of the work in ensuring that questions related to the accuracy or integrity of any part of the work are investigated appropriately and resolved.

http://www.icmje.org/recommendations/browse/roles-and-responsibilities/defining-the-role-of-authors-and-contributors.html

---

Publishing Ethics and Research Integrity

The authorship guidelines have been established to combat problems that have appeared in scientific publishing. These include ghost authors (people writing papers but not being identified as authors) and gift authors (adding people to author lists who didn't contribute). Both scenarios should be stopped.

- **NEVER** submit a manuscript for publication without prior approval from your co-authors. That approval should be in writing so that you have a permanent record.

- **DO NOT** accept authorship status on a manuscript where you have not contributed or are not happy with the submitted version. Any problems with the paper in terms of research integrity will severely negatively affect your career.

Acknowledgements

The acknowledgements section usually goes near the end of the manuscript. Here you can thank other people or entities who have contributed to your work. This includes the funding bodies that supported your work and you should note your appreciation to patients who provided time or samples where appropriate.

If someone does not meet the authorship requirements but still helped the paper in some way, then you can indicate their precise contribution in the acknowledgements section. For example, if you have someone who contributed patient samples or reagents, or who helped out in the lab but didn't contribute to the writing process, or if you used technical assistance in a core facility, then you would thank them in your acknowledgements. Similarly, if someone helped edit a draft of your final manuscript for grammar or English usage but not for other aspects of the scientific interpretation, then they likely do not meet authorship guidelines but you should credit them in the acknowledgements.

I see you were a co-author on this paper

Can you explain the choice of statistical approach used in figure 4?

Also, the experimental methods used

Don't be surprised if you get asked in interviews about papers where you are a middle author. You should be able to explain and defend any published work with your name on it!

Recently, some journals have adopted a style of acknowledgement where you also directly indicate the contribution each author made to the manuscript. This is an attempt to make everything more transparent and account for cultural differences in publishing practices. Even if the journal doesn't require it, it still can be good to include something such as, "KH designed, carried out the experiments and drafted the manuscript, CS conceived the study, analysed the data, prepared the figures and drafted the manuscript, GZ analysed the data and drafted the manuscript" etc. Note that the journal might define the phrases to use.

7.3 Choosing an Effective Title

How do *you* find a paper or decide which abstracts to read? The answer almost certainly is the title. Getting the title right will help your work get noticed, get read and ultimately get cited. Get it wrong, and it will mean that the work you did does not get seen by as many people and, consequently, will be cited less frequently. Research that no-one knows about might as well not exist. However, there is a pitfall, a title that is not supported by the data will mean that the manuscript gets rejected during the review process or, at very least, have major modifications requested.

The advice here refers primarily situations where you have collected new data or performed secondary analysis such as meta-analyses or systematic review, e.g. research projects or data manuscripts for publication. Titles for other types of science writing such as review articles, essays, or opinion pieces are a little different as you have more flexibility to use humour, puns etc. but for research reporting, you almost always should be serious, direct, and unambiguous.

Make sure you check the rules! Each journal has a different maximum character or word count.

Draft a title near the start of the writing process

You need to know what you are writing about, the story that you will be telling. Therefore, before you get into the meat of the introduction and discussion, you need decide what you consider to be the *primary conclusion* from your paper and write a draft title that states that message as simply as possible. Think about what part of the study is the most valuable to the field and write that as a statement that captures that point as simply and comprehensively as possible.

Big Tip

You need to know what your story is about before you can write the story!

The draft title doesn't need to be word perfect, but your figure order, your results, your introduction and discussion all depend on what the message you are trying to deliver. Without a rough title in place, you will be vague and unfocused. Having something down on paper will help you keep the rest of your writing "on message".

It's never too soon to start thinking about the title. Indeed, you might have already thought of it; the overarching hypothesis of your project might turn out to be a decent title. I'd go one step further and say that coming up with a draft title for a manuscript is something useful to do while still gathering data. Knowing what you think the story will become, is a good way to prioritise your next set of experiments. Your title, of course, may need to change based on any new data obtained, but working toward a clear goal is a very effective way to focus your efforts and to identify when you are ready to write.

With a draft title in place, you can go ahead and make your figures write your results, intro and discussion. Come back to the rest of this section again when those sections have been drafted and are ready to edit the title to increase its impact.

Styles of Titles

These are the four main types of titles; I've ordered them here from best to worst in terms of effectiveness:

Types of titles

[Best] Conclusions Statement (aka declarative)
"LaNt influence laminin organisation."
Compound (statement + question, or statement + implication)
"LaNt influence laminin organisation: implications for wound repair."
Descriptive
"Investigation into the effect of LaNt on laminin organisation."
[Worst] Question (aka interrogative)
"Do LaNts influence laminin organisation?"

I've rated declarative conclusion statements as the best as they are the most useful to the reader. A statement title tells the reader exactly what they need to know without having to go any further. In contrast, teaser style questions are annoying! Clickbait might work for getting people to open the abstract, but you'll miss out on citations in the longer term. More generally, I think a choosing not to make an impact when you can, is wasting an opportunity to shape your readers' thought processes.

If you are writing a project report for an assignment, then a definitive results statement titles will lead to better marks. When you use a question or descriptive title, then the *first* thing your marker will think is that you don't know what your data means. Even if the rest of your work is super clear, the very first impression they will have will be negative. Compare that with a statement title that tells your marker that you do understand and grasp the significance of your work; a much better first impression.

In a thesis or dissertation, I recommend giving results chapters statement titles too. You could call them Chapter 1, 2, 3 or use descriptive titles, but once again, a declarative statement will more rapidly establish the value of your findings and help you to shape your narrative.

What works for journal articles?

Putting the general style to one side, there are some additional points to consider:

General Advice for Titles

- Shorter statements work best.
- One clear message is better than two less clear ones; focus on the most important finding.
- Think big picture (within reason). Write your title to appeal not only to the target journal but also to as wide an audience as is appropriate.
- If you need subclauses, put the primary finding clause first. "LaNts regulate laminin deposition by corneal epithelial cells" is better than "In corneal epithelial cells, LaNts regulate laminin deposition."
- Using the active voice sounds better than passive. For example, "LaNts regulate laminins" sounds better than "Laminins are regulated by LaNts." But be careful! Those two sentences might mean different things, your data may not support both options.

Where does this advice come from? Well, there has been quite a lot research on which titles work, in terms of downloads and citations. Some of this is subject-specific, but here are some clear trends:

- Papers whose titles emphasise broader conceptual or comparative issues fare better both pre- and post-publication than papers with organism-specific titles.
- Articles with question titles tend to be downloaded more **but cited less**.
- Articles with longer titles are downloaded slightly less frequently than the articles with shorter titles, and articles with shorter titles tend to receive more citations.
- Titles with a colon tend to receive fewer downloads and

> **Big Tip**
> Papers with statement titles get better marks or get cited more frequently than the exact same work with a question title.

citations (and will be longer) than those without. An explanation here is that the secondary clause, after the colon, is often a caveat, qualifier or limitation, so this final point is likely to be reflection of the breadth and implications of the study rather than necessarily a feature of the title.

Your paper will be remembered and cited more frequently if you have a simple, clear statement. People will see it in their reference library, remember what you discovered and therefore reference you.

Use the best title for *your* story!

The citation statistics lead to pretty clear advice; *short definitive statements are best.* But, be aware that the correlations for each of the findings are not super tight, and there are quite a lot of confounding variables which influence the data interpretation.

Like all statistics, these data tell you about the population, rather than specifically about any individual piece of work. Importantly, they don't say what will definitively work for *your* story. My message, therefore, is don't try to shoehorn your title into a framework based solely on these findings. This same point holds true for much of what will follow in this chapter, it is advice rather than rules. You will have to make some decisions along the way (with your co-authors help, of course). For example, recently my group and I were writing a paper, and even though we tried to shape the title into a statement in different ways, in the end we were unable to come up

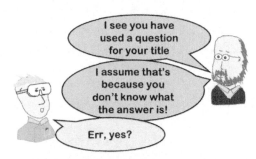

The best titles are clear statements of what your data means

with something that effectively captured the message and which was succinct and effective. We published that work with a descriptive title, and although I was a little unhappy about this, it was the right title for that body of work.

Write lots of versions of your title

I usually come up with a list of options and then ask people who are unfamiliar with the findings to help select the one that is clearest in its message. Also get people who *are* familiar with you work to check that you are stating the findings appropriately, not over or understating them, that the phrasing is unambiguous and that you haven't (accidentally) used a question! Titles are important, and it is always worth getting feedback.

What to check when editing

Don't undersell but also don't overstate your findings!

If you legitimately have cured cancer, then you should shout it from the rooftops. However, if you cannot support such a bold statement, then you should not say so in your title! Ok, that's an extreme example, but the concept stays true; your title should capture the biggest finding of your data but only go as far as your data can support. Scientists need proof. A reviewer, editor or marker will spot the overstatement, and they will ask you to provide the proof (or will reject your manuscript).

Generalisability

If you performed the work in a limited capacity and you do not yet know if the findings translate to other systems, then you need to indicate that limitation. Failing to do that is a form of overstatement that reviewers will not accept, and readers will not like. For example; "LaNts regulate cell migration" sounds bold and dramatic, but if the data were only from corneal epithelial cells, then it would be inappropriate to imply the effects can be generalised to *all* cell migration in all cellular and tissue contexts.

Presenting implications as concrete findings

If your data shows that treatment with drug A decreases protein levels of B and drug A also changes cell behaviour C, *but you haven't shown that C depends on B*, then you should not claim that it does so in your title. Again, this is a form of overstatement, and it should lead to your reviewers rejecting or requesting revision of your manuscript. Sounds obvious, but it is frequently seen.

Danger words: watch out for these

Via –have you really connected the two findings in the way you are stating? For example, "miRNA 485 regulates differentiation *via* downregulating pax3", sounds OK, but miRNAs regulate many transcripts. Therefore, unless there are data from conditions where miRNA 485 treatment was unable to regulate pax3 but could still regulate everything else, then the *via* would not have been proven, you are in danger of overstretching your findings.

To – implies a deliberate mechanism. Watch for anthropomorphism; assigning character traits to things like cells or proteins, which don't exhibit thought. Using *to* as shorthand for *a pathway has evolved to* or equivalent, is quite common but there might be more elegant and less ambiguous ways to make the same point (simply switching *to* for *which* might tighten it up).

Ambiguous phrasing

Be very careful about whether your title could be read in a different way. Even if it seems obvious to you, any uncertainty about how the phrase is connected is a bad thing.

Practice

An exercise that I encourage my students to do, especially at journal clubs, is to critically evaluate the title of a paper that they have just read. The more you do this, the more you will realise that all the mistakes I have listed above do happen and that they do so with depressing regularity. Hopefully, by training yourself to recognise mistakes in others, you will become better at avoiding those same pitfalls in your own work.

7.4 Scientific Abstracts

An abstract is not so much a teaser, but rather a short overview of the whole paper, the complete story told in fewer words. Your abstract needs to capture the question you asked, the motivation that drove you to ask that question, the methods you employed, the results you obtained, and what conclusions you can draw from those results. It's a lot of content, and you have very few words to do it in.

The abstract is a mini version of the whole paper:
Motivation, Aim, Method, Results, Conclusions

First impressions matter

Every part of your paper matters. However, the abstract is the first time your reader will encounter your writing and learn about your story. It is the quality of the abstracts that makes the difference between a person choosing to read the rest of your paper or dismissing it as uninteresting. The abstract certainly makes a difference about whether your paper gets cited or not.

Like the rest of your document, your abstract will need multiple drafts before it is perfect. Some people say to write abstracts last, while I appreciate that sentiment, I find it helps to have a really rough draft in place early in the process. Write the abstract can help you to have

Big Tip

Always ask a non-expert to read your abstract before you submit.

cohesion in the message running throughout the story Irrespective when I draft the abstract, it will get read every single time I open the paper. The abstract becomes the part that I spend the most time getting right.

It is always worth getting someone who is completely naïve to your study to have a look at your abstract. Abstracts should be able to stand alone and, because they are quite short, most people will willingly do this for you (and you should be willing to do it for others). Ask your peers if the abstract makes sense. Do they understand everything? Would they want to go on to read the rest of the paper?

Structured or Freeform?

Many journals are prescriptive in the layout of their abstracts, requiring specific subsections i.e. *structure*. The subsections are some variant of background or introduction, aims and objectives or hypothesis, methods, results, and conclusions. Each of these sections gets one or two sentences, with the results part being a little bit longer than the other, maybe three or four sentences.

Freeform abstracts, in contrast, take on more of a narrative, storytelling style. Overall, you should follow the same core structure, but you have more flexibility to tweak the order to improve the rhythm and make your abstract more enjoyable to read.

The most obvious way to take advantage of the freeform option is to combine the methods and results for each stage of the study and to use linking phrases to smooth your transitions between the stages of the project. For example, phrases like "next we asked" can serve useful links to transition into the next piece of work.

Structured abstracts are usually easier to write. Removing the option for creative flexibility also takes away some of the decisions that you as the writer will need to make. You cannot go too far wrong. Indeed, it can be a good idea to use the subsections from a structured abstract to put together a first draft irrespective of whether they are needed or not. You can always remove the subheadings at the end and add some link phrases to make it a bit less dry.

Length

Check the word or character count limits for your target journal. These usually somewhere between 200 and 350 words but can vary a lot from journal to journal. In the rest of the writing sections in this chapter, I won't focus much on length as, if you follow the guide layout, you usually reach approximately the right length. If your first draft ends up too long, it is not necessarily bad as it is easy to reduce the word count and those changes usually improve the manuscript. However, the difference between a 200 and a 350-word abstract is quite large in terms of style and content. Therefore, I do recommend that you make yourself aware of your target length before you start and then aim to be within 50 words either side of that limit in your first draft. If there are no specific requirements or you don't yet know the target journal, then 300 words is a good starting point.

The box below has some suggestions on balance. These values can vary quite a lot depending on your story, but, again, are a good guide place to start.

Recommended proportion of word counts	
Motivation and Background	~20%
Hypothesis and Aims	~10%
Methods	~20-30%
Results	~30-40%
Conclusions	~10%

Getting started with your abstract

One way of writing the first draft of an abstract is to think of each sentence as the answer to a question.

Sentence	Question
1	Why should the reader care about the topic?
2	What key things does the reader need to know to understand your study?
3	What specific thing was **not** known about your research area before you started? What question did you address?
4, 5	How did you do your experiments?
6, 7, 8, 9	What results did you get?
10	What do your findings mean?

Introduction: Context, Background and Motivation

Your first sentence should be about the *big picture* and should be accessible to a scientist from any discipline, not specifically yours. Start reasonably wide but on message. You are telling the reader why they should care about *the topic*.

Follow the big picture sentence with one sentence of more detailed background in which you set up your specific question. Here your goal is to tell the reader what *wasn't* known about your research area before you started and what value you are providing by filling that research gap. Remember that just because something is new doesn't mean it intrinsically has value, your sentence should *emphasise why it is worth knowing this new thing.*

Phrases to establish value

It is not yet known...
It has yet to be established...
Little is known....
... has never been confirmed.
It remains controversial...
This raises the question...

Question: Aim or hypothesis.

Use one sentence to state, in as clear a way as possible, the problem that *your* work addressed. This sentence can be framed as an aim, objective or hypothesis. It works best if you construct the phrase as a single entity that all your experiments point toward, rather than as a series of questions. If your goalposts moved as your work progressed and you followed the science, then write the aim sentence as the starting point of what became the final story.

What's the difference between a hypothesis and an aim? Hypotheses are discrete testable statements; an experimental series will test hypotheses. The results of the experiments will either cause the hypothesis to be rejected or will support it. In contrast, aims are what you want to find out. You do not necessarily need to write the words aim or hypothesis in your abstract. Be aware, your instructions or the journal may define what style of phrasing they want you to use.

Aims vs Hypotheses

Hypothesis: "Treatment X reduces Y."
Aim: "Determine if X reduces Y."

Methods and Results

If you are writing a structured abstract, the results and methods sections may be broken up into two distinct subheadings. If so, the methods can be short and to the point, "X was investigated using ...". If using a narrative style, you can combine methods and results into one; "using (the approach) we determined that (the results).

For most papers, the results should be the longest part. Summarise your main findings, not necessarily everything, focusing on the parts of the story that answer the main study question. Indicate the magnitude of effects, the strength of your inferences

---

(i.e. p values) and the statistical tests used. In other words, you should include actual results in your abstract. Draw attention to anything unexpected and indicate where your data support or refute your initial hypothesis. The results will read better if you can tie the individual findings together with link phrases so that it flows into a coherent narrative.

Conclusions: "These findings have implications for..."

> **Big Tip**
>
> Don't only summarise your findings. Define why your results are interesting and important, and connect them to the wider world.

Your goal at the end of the abstract is to put your work back into a real-world context. Make your conclusions as specific and as generalisable as your data can support while making sure they aren't an overreach. Your main conclusions should reflect your title (assuming you used a statement), and later in your discussion you will expand this conclusion statement into a fleshed-out paragraph to complete the manuscript. Use different phrasing between the abstract, discussion and title but make sure all three deliver the same core message.

The conclusion as a proportion to the whole word count depends on how big, complex or impactful your findings are. Usually one or two sentences is all you will need. Some journals have different requirements; for example, Nature expect the conclusions section to be the biggest part of your abstract; high impact journals are all about the message and wow factor!

References

Usually, *don't* use references in an abstract. The only time you might have to do so is if your whole story is derived from a single previous finding. On the rare occasions when you must cite something, you should be aware that there will be different rules for how to format the citation in the abstract than in the rest of your document. You need to provide a near-complete citation including titles, authors and page numbers within the abstract. Abstract citations use up a lot of your precious word count!

Use your conclusion sentence to put your work back into context with the real world.

Keywords and search engine optimisation

It's all very well writing a great abstract, but if no one reads the abstract in the first place, then you are no better off. Your title will do most of the work in attracting reader, but you still need your paper to appear in the results whenever people are performing literature searches. The abstract should contain the keywords people will use when searching. Test out some search terms and see what comes up.

Search algorithms keep improving and changing how they work; however, here are some general tips you could consider:

- Place your main keywords in the first two sentences
- Be consistent in terminology throughout your manuscript, including subheadings.
- Don't overuse keywords; too much repetition ('keyword stuffing') may result in search engines 'un-indexing' your article.

For biomedical work, you can use resources like the library thesaurus (National Library of Medicines) to help you find effective keywords. Google AdWords keyword planner or Google trends can also be good sources of information. The keywords you use in the abstract are also the keywords you should provide on your title page.

Big Tip

Don't let the quest to include keywords distract from the delivery of your story!

Note: don't go too far in your edits, keywords are important but certainly you should not compromise on the reader experience for the sake of search engine optimisation!

Thesis or dissertation abstracts

The concept of an abstract for a PhD thesis or Masters dissertation is fundamentally the same as for a paper, except you will have a lot more work to cover and a bigger word count to play with. You should look at the abstract as a way to tell your examiners which parts of the larger study are the most important. It can set the scene for all the work to follow and is a good chance to highlight the value, extent and novelty of your work as a whole. Whenever I am preparing to examine a PhD student, I read the abstract before I read the thesis and then again just before the oral exam (*viva voce*). This second reading is useful for me to recontextualise all the results.

As always, step one is to check the instructions. If there are no specific word limits, then I recommend using the equivalent one single-spaced page, about 700 words, if you find you are needing longer than that it might be a sign that you haven't been very discerning about what you have selected to highlight. You don't want to frustrate your examiners before they have even seen a single data point.

In terms of structure, it is quite similar to a manuscript abstract. Long-form abstracts are almost always freeform, so you have flexibility to do whatever works.

As usual, start with the motivation and rationale, then a clear statement of the problem that your work addressed. Limit this to about 20% of the words, usually one short paragraph. For methods and results, you have two main options. If you used a single model system or a limited range of approaches throughout the work then describing the experimental system using its own paragraph might make sense. This can work, for example, if you have generated a new transgenic mouse model or a genome-edited line, and then characterised it in the rest of the thesis. If instead, you have results chapters that less obviously connected to a central theme, then an easy option is to use one paragraph per chapter with a brief overview of the methods and major results presented together in each paragraph. Altogether, in whichever format you decide, the combine methods and results should make up about 60% of the abstract. The final 20% should be dedicated to conclusions. This might sound like a lot, but you should cover what the findings mean, whether they answer the problem statement, and how they have advanced the field.

I don't recommend including future directions in the abstract, you don't need anything to distract from what you have done. End with a sentence that wraps up the story by saying what your work means in relation to the original motivation. Make this upbeat in tone and emphasising value.

What to check when editing

Balance

The most common problems I encounter in new writers' abstracts are to do with balance between the different elements. Mostly the issues are that the early drafts include too much background information and therefore don't have enough words left to give enough attention to findings. You probably don't need as much introductory material as you think! When editing your draft, aim to cut down any waffle and get to the meat of the story as soon as possible.

Conclusions

Your conclusion is essential, don't forget it! It is required to round out your abstract and make it a complete stand-alone entity. If word count is tight you have to choose between one extra results sentence and including a conclusion, always pick the conclusion!

7.5   Introductions to Research Papers

The introduction to your research paper sets the tone for the rest of the work. It does an important job. It tells the reader, marker or reviewer the motivation, *rationale* and the necessary *contextual information* they need to understand your study. If you do it well, the introduction will make your readers eager to read the results and discussion.

Paper vs Thesis

The material in this section is geared toward introductions to short format writing (i.e. research papers).

In a thesis or dissertation, you might have multiple introductions. An introduction to the thesis, an introduction to the literature review and

Big Tip

This is the introduction to your story.

It is **not** supposed to be a recap of the whole field.

then introductions to each results chapter. For the main thesis introduction your job is set the scene for the study and then give an overview of where you are taking the reader. Its role is primarily to act as a signpost. Much the same in a lit review, here you usually only need one paragraph which contains forward references to the different subsections and why they are needed. "Later in this thesis I will describe … to put that work into context I will first discuss … and …".

For the results chapters, you may or may not need a specific introduction, it depends on how your overall thesis is structured. However, if you do, then using a similar format to a manuscript (described below) works very well. You don't have space limitations in the same way that a manuscript does, but the advice is still to stay as focused as possible.

Objectives of an Introduction

You might think that an introduction is simply to tell the reader all the information needed to be able to understand your work. While this is true in a fundamental sense, it is often better to think about an introduction as having four discrete objectives. Thinking about an intro this way will help make sure you hit all the points you need to maximise value.

1.   **Motivation:** establish the wider context of your work and why it matters to the world. This should be framed in a way that the reader will care about your work.

2.   **Mapping the Gap:** define what *wasn't* known before you began and why it was worth addressing that gap. Mapping the gap should lead into making a *problem statement*. This is simply a sentence where you should define the problem your work addressed. Don't map a gap of something that your study didn't actually address! Those gaps can be brought in as next steps in your discussion section.

3.   **The rationale for your hypothesis:** provide enough detail from previous studies to make it clear why you formed the hypothesis that you went on to test and/or the basis of the approach you took. Ideally, the lead up to a

hypothesis statement should be such that a reader would also form the same hypothesis given the same background information.

4. **Basis for discussion:** lay the foundation for any points that you will bring up again in your discussion. This final point is why it is best to plan the introduction and discussion together and then later edit them alongside one another. A question you ask or a gap you identify in the introduction should always be answered or addressed in the discussion.

> The biggest difference between a good introduction and an OK one is that the OK introduction tells the reader things that are known and are relevant but doesn't really make it explicit why the reader needs that information. In contrast, a good introduction frames all the information in a way that makes it clear how the infomation connects to the study. It highlights what *wasn't known,* and establishes why knowing the answers would be valuable. A good introduction helps the reader to see where they are going. Signposting and forward references are important in achieving this goal.

Plan the main story of the paper before you start

Before you can write an effective introduction, you need to have a clear idea of the main story of your manuscript. In essence; what is it that you are introducing? Without a clear direction, it is very likely that you will write an unfocused and, almost always, overly long introduction. It is for this reason that it is usually best to write the introduction after you have drafted the results and/or abstract.

While the advice of writing the introduction near the end is undoubtedly good, I have supervised enough students to realise that everybody wants to get started as soon as possible and writing a draft introduction seems like a logical thing you can do before you even any data. Let's face it, the information that set up the study won't change. The problem is when you have spent time writing an unfocused draft, it can be hard to have the required discipline to delete whole sections that you later decide aren't relevant.

So, what can you do? Well, as you will have a good idea of the broad topic area you can do the required background reading and take notes of the important things you want to introduce or that you think you will

Make sure everything in your introduction is directly relevant to the studies that you have performed

discuss. You should be reading a lot at this point and not of it will be directly relevant. Therefore, while reading and note-taking, try to identify how each paper connects to your study question or to the data interpretation (if it will appear in the intro or discussion).

Once you have read everything, your next step is to sketch out a framework for what you think the contents of your introduction will need to be. Later you can come back to fleshing out the text once the rest of the pieces of the story are in place. As always, writing involves drafts and then editing. Adopting this approach doesn't mean things will not need to be deleted, but hopefully you won't have as much off-topic material if you focus on getting the story right first.

Introduction Structure

> ## Recommendations
> ~1 single-spaced page = approx 500-800 words.
> 3-5 paragraphs.
> No subheadings.

The standard way to organise an introduction is to start wide and become more focused and more specific as you progress. When deciding exactly how wide to start, think about the journal's target audience. The introduction for a multi-discipline journal will need to start wider than a manuscript aimed at a more specialised area. Note that the biggest problem new writers tend to have is that they make the introduction too long by either starting too wide or going into too much detail in non-essential areas. Make sure you stay relevant by asking yourself *would it matter if I didn't include this sentence* and then delete anything you don't need.

Paragraph structure is really important too! Be aware that established researchers in the field will already know most, if not all, of the content in your introduction. These readers are likely to skim read your intro and progress rapidly to the main story. Skim readers focus one the first and last line of each paragraph, they absorb that point then move on assuming that they know the information in the middle. This point means that you should make sure the *topic* and *wrap* sentences in each paragraph work to advance the story and are sufficient to set the scene and lay foundations for the future.

Paragraph One- Motivation and Big Question

The opening sentence

You need to be on message right from the start or your work will feel unfocused. Open your introduction with a big impact statement.

Ideally the first sentence of your first paragraph should include one of the main words from the title. Starting with one of your key words should stop you going too wide and thereby avoid adding an extra paragraph of unnecessary information

Don't start your introduction too wide. Use a keyword from your title in your first sentence to help you focus

In the biological sciences, our work often starts with a disease or environmental issue. As a general rule, if your title mentions the disease/issue then your opening statement can talk about the scale or severity of the problem. These are relatively easy to write in a way that connects with the reader. However, if the work you have done is more mechanistic, less directly related to the disease but rather is about understanding the underpinning biological basis, then your opening should introduce the biological structure or protein pathway etc., in the first sentence then come back to the relevance of the work to the disease later on.

Big Tip	
Include a keyword from your title in the first sentence.	

Example first sentences

Paper on tendon stem cells.
Tendon is prone to injury and degeneration, and this is often seen in occupational and sporting environments.
Lee, K.J., Clegg, P.D., Comerford, E.J. et al. BMC Musculoskelet Disord (2018) 19: 116.

Paper on a basement membrane protein.
The alveolar compartment of the lung contains a unique basement membrane, which is shared between epithelial and endothelial cells.
Urich, D., Eisenberg, J.L., Hamill, K.J. et al. J Cell Sci (2011)124(17): 2927-2937.

Paper on genetic causes of atopic dermatitis.
Atopic dermatitis (eczema) is a common inflammatory skin disease affecting 15-30% of children and 5-10% of adults.
Paternoster L., Standl, M., Waage, J., et al. Nat Genet (2015) 47(12) 149-1456

Have a look at some opening sentences in papers in your discipline, and you should see a similar trend.

Paragraph One - sentences 2-4

The next two or three sentences of the first paragraph should expand on the opening point to establish the *motivation* behind the work. Usually you can do this by making big-picture comments about the real-world problem and/or where the lack of knowledge lies. For example, most introductions about a disease will mention the effect it has on people's quality of life, how many people are affected, and what it costs to treat. They will then close by pointing out where there are challenges, such as a lack of effective therapy or an area where there is incomplete understanding.

Many of the statements in these early sentences will be quite general, so it will often be appropriate to cite a review articles rather than requiring primary references. Ideally you should reference work from different authors, including competitors and collaborators, to give a broader coverage of the existing literature. Of course, any very specific comments will need the appropriate primary data reference, and you should not cite a review if the point you are making comes from a single original source.

Paragraph One - wrap sentence

The end of the first paragraph should tell the reader where you are going to take them.

The last sentence should say something like: "In this study, we...". You don't need to use those exact words,

Big Tip
In the final sentence of first paragraph point out a problem or hole in the knowledge that your research paper will address.

but the last line should capture that sentiment. Note that you are not writing deep detail here, that comes later, you just need a one sentence overview of the subject area to focus the work and allow you to progress to paragraph two.

Paragraphs Two, Three and Four

Now that you've established the big picture, the overarching goal of the project, you can start to go deeper, moving toward the specific questions your study addressed.

Before you begin, write down what you need the reader to know. This is not the same list of *everything* that you know, but rather the information that is relevant to your current story. These middle paragraphs are also where you drop in the details that will refer to again in the discussion, so identify what set up you need for your discussion too. Put your list in order of importance, then take the most important points and turn them into the topic sentences. Remember, topic sentences should be accessible and mostly do not need to rely too heavily on specific pieces of information. Writing the topics early is so that you don't lose focus. You don't want to go off track here, if a point is merely an interesting aside it probably has no place in your introduction.

Next, order your topic sentences from broad to narrow. You should constantly build toward the goal of establishing your specific, narrow research question. Once you have ordered the topics, all you need to do is use this framework to assemble the rest of the story. Most of the citeable material, the info that has come from preceding studies, will appear in the body sentences of the paragraphs. Those should ordered to connect with an appropriate topic sentence.

Throughout these paragraphs continue to "map the gaps" that were addressed by your work and highlight why they matter. These gaps though should be much more specific, literally the things that you actually answered. By the end of this section, your research question should be obvious and usually you should use the last sentence of the penultimate paragraph to state that question.

Remember that you want your readers to be interested in the answer to the questions you have asked. If you have not yet reached that point, then you need to tidy up, rephrase, reorder and possibly cut out non-relevant material. I tend to overwrite my first draft of my introduction section then cut out or contract lots of the sentences during the editing phases until I get to a stage where I am happy.

Example phrasing to "Map the Gap."

- "Previously, it has been reported that X affects Y, *but the molecular mechanism is unknown.*"
- "An increase in Y has been correlated with disease progression, *but the pathogenic consequences have never been established.*"
- "Protein Y has been established as important for cell-type B function Z, but its role in cell type C *has not been evaluated.*"

The phrases in italics are examples of how you can draw attention to where your research is going. These sorts of phrases are most effective in the topic or wrap sentences of a paragraph.

You will need numerous primary references throughout the middle paragraphs, usually more than one citation per sentence. Again, make sure to cite your competitors, not just your boss, and mention any area where there is controversy. The controversy point is especially important if the data you are about to present supports or contradicts previous findings. Your reviewers will know the field, so do not ignore some previous work because it doesn't fit your model!

Frame your introduction so that it focuses on what *wasn't* known rather than just saying the current thinking

Final paragraph

By the end of your set up paragraphs, the research question should be implicit. In the final paragraph, you should now make it explicit. Clearly state the question, the hypothesis or objective that you have been working towards (only use one of these options). Simple phrasing like; "The purpose of this study was to determine....", or even just; "here we tested...", can be very effective.

Follow this opening statement with a few sentences explaining your experimental model system or an overview of your approach ".... using X, Y and Z". You may want to say why this approach was used if it is a new approach. It is quite common to end the introduction with a one-sentence overview of the results and what they mean in relation to the big picture question you posed in the first paragraph of the intro.

Big Tip

Shorter introductions are usually better.
Make sure everything is relevant and on message.

And that's it. You have written your first draft. Later in the writing process, you'll come back and edit to make sure every sentence is helping your story and to cut out everything that doesn't add value. Introductions end up being one of the easier parts of a paper to write, but, like all things, it will take practice. Don't worry if your first draft comes back with lots of comments from your co-authors.

What to check when editing

Have you written for your audience?

Think about who you are writing for and how they will absorb your work. This is true when editing all sections, of course! Here you should check if you have started too wide or too narrow for your target audience? Have you used language that your audience will be comfortable with? Writing in an accessible way is good for everybody, you shouldn't be aiming to make your work dense merely to try to sound clever.

Check your paragraph structure

Look at each of your paragraphs and make sure that they are complete and structured appropriately with all the required elements; topic, token, links, and wraps. Pay special attention to any paragraph that is below ~100 or over ~200 words in length as they are often the ones with problems.

Read the first sentence of each paragraph. On their own, these sentences should be enough to set up the main points of the story. Each paragraph's topic sentence should also look *forward* toward the contents of the current paragraph rather than backward.

> **Big Tip**
> If someone were to read only the topic sentences, they should be able to grasp the basis of the whole story.

Next pay extra attention to the last sentence of each paragraph (the wrap), make sure that every paragraph has one, and that each wrap effectively advances the story on to the next paragraph. If you ever receive comments about a draft "lacking flow" then usually it is a sign that your wrap sentences aren't pushing the story forward.

Remove any off-topic material

Being self-critical is one of the hardest parts of writing, but the key to writing well is to cut out anything that should not be there, irrespective of how much time you spent writing a section. For every "fact" you are delivering, ask yourself what value it adds to the story, if it isn't adding anything then delete it.

I commit this paragraph to the cemetery of irrelevance

It can be hard to delete work that you spent a long time reading and writing to prepare, but, if it isn't relevant, it has to go.

The parts of your text where the answer is clear, where the work is either definitively useful or not useful are easiest to fix. It gets harder when you think everything is needed. In these cases, look at the relative value to the story of each part and then devote space proportionate to their importance.

Referencing

Make sure every piece of information which is not common knowledge or an original thought is properly cited. Over citing is better than plagiarism!

7.6 Materials and Methods

The *Materials and Methods* also known as *Experimental Procedures* section should be the easiest part to write of any scientific report, manuscript, thesis, or dissertation; after all, you know what you did! Having said that, there are a lot of common mistakes that come up in new writers' early drafts but the good news is that these mistakes are easy to fix, and that's what this section is about!

Keep a good lab book

Having a clear, understandable, and complete lab book makes everything easier. Your lab book should have all the details, including manufacturer information, product codes, lot numbers etc. The more details you have recorded, the easier and quicker this section will be to write. Remember that your lab book should not leave the lab. You will need to scan or copy pages as required if you will be writing at home.

Write as you work

Whereas for most of the rest of your manuscript, you need to already have a solid idea of what your main story will be before you really can get going, the methods sections are different. Here you can and should write everything as and when you are doing the actual experiments. There are two simple reasons for this:

Write your methods as you work!

The best time to write your methods section is as soon as possible!!

1. **You will find it easier.** You will remember exactly what you have done and will not have to rely solely on the quality of the records in your lab book.
2. **You need the details.** If you have forgotten to record the batch or clone number of some key reagent it is much easier to just look in the fridge for the details if you write as you go.

Use published examples as a guide

For most experiments, your protocol is likely to be based upon and very similar to published work. So, use the publications to help you. This is especially true when your supervisor has published the same technique before. You shouldn't be copying, but rather rewriting while carefully checking and updating the details of anywhere you made changes.

As a member of the next generation of scientist, you should be aiming to do things better than the generation before. Wherever you can improve clarity or specificity in your methods, you should do so. Remove ambiguities, tighten up phrasing and add experimental specifics missing from the original. Pay extra attention to anything required to *interpret* the data. As always, do not forget to provide the references to your source materials including previous publications of methods.

If the core aspects of your experimental protocol are identical to those published before, it's OK (in a journal article-type paper) to provide only the link to primary reference, "X was performed as described in …". However, whenever you use a reference *instead* of writing out a part in full, make sure the paper you cite contains the complete methods. You should not send your reader down a rabbit hole of searching for information. Also, check that the cited paper is freely available online. If it is behind a paywall, then you should write an abbreviated method so that *all* of your readers have access to the information.

Serial referencing is a common problem!
If the material you are referencing isn't in an open access journal make sure you provide enough detail to complete the experiment in your manuscript

Thesis writing

Most methods rules are the same for a thesis as for a paper. The only main difference is that in a thesis or dissertation, you do not have tight space limitations. Therefore, I would expect to see the full experimental details written out as well as the relevant citations to where the original method came from.

General Rules for Methods Sections

Writing

- Past tense.
- Third person (do not use I or we).
- Write in prose, using complete sentences and proper paragraphs.
- Do not use bullet points or lists.

Units

- Use the SI form of units / measurements.
- Use a space between the number and its unit (1 mm)
- Do not use a full-stop or period "." after the unit.
- Use concentrations rather than mass and volumes whenever you can.
- Not sure which units are correct? Check the ukma-style-guide

Information

- Include **all** the details required to perform and interpret your experiment.
- Avoid repetition.
- Reference where protocols came from and where reagents, cell lines or approaches were validated.
- Include details about ethical approval wherever appropriate, usually in its own subsection.

246

Materials

- Provide supplier information for all reagents, including gifts from people.
- Include the full name, city, state and country at first mention (LaNt Inc., Dundee, UK) then the company name only at subsequent mentions (LaNt Inc.).

Data analysis

- Include full details of analytical techniques.
- Define your statistical approaches fully (more on this below).
- Do not talk about any of the data. The data go in the results.

The Details Matter

The purpose of this section is not only so that other people can repeat your experiments. It is also to enable your readers to critically evaluate what you have done. Your readers and reviewers will want to identify how robust your data are and where there are any limitations. It is the details of the experiment are what make the difference in whether this is possible or not. At every step along the way you have decided on what or how to do something or what reagents to use, these decisions influence the data and they need to be clear for all to see.

Importantly, the methods section needs to include everything before data collection *and* everything after. Knowing that your analysis was approached in the correct way, that you chose the right sample size, defined independence appropriately, and used the right statistical tests are all key to determining if your conclusions are justified.

If you don't write your methods effectively and comprehensively enough, then your reader, reviewer or marker will have to fill in the blanks Scientists are a critical bunch, we are trained to take data and make inferences, and if you choose to leave out details, then we will assume there is a reason for not telling us something!

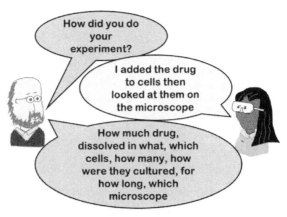

Tell the reader *everything* they would need to repeat your experiments. The details matter.

In case you aren't already convinced, think about when you come to defending your work during your talk or viva. Any place in your methods where you have been incomplete or ambiguous, you are likely to be asked questions. Will you be able to remember the details of a series of experiments you did months or years ago while already feeling the pressure of your exam?

Structure

Separate each complete experiment into its own subsection

Help your readers locate the material they need by breaking your methods section into headed subsections. Each subsection should be big enough to cover a whole experiment rather. If you find yourself writing just one or two lines per section, it is probably too short, look for ways to combine those sections. Conversely, if you have with more than four paragraphs in one subsection, then it is probably too long and might be time to split. Subsection headings should usually be descriptive and based around the approach; "RT-qPCR", "cell culture" etc.

Write the big, general approaches first

Your goal is to deliver your methods in as clear and succinct a way as possible. Repetition is your enemy!

Start with the parts that relate to multiple aspects of your study. Things like descriptions of antibodies used, cell lines and culture conditions, patient recruitment, mouse lines, or construct generation. If you can establish common points first, then you will only need to mention them once. For example, if you used the same plating or treatment strategy for your cells across multiple different experiments, then describe the plating strategy once in a general section at the beginning with a lead in such as "for all cell-based experiments...". If you used multiple lines for the same experiment, describe the consistent things once, then highlight the specific details at subsequent mentions.

> **Big Tip**
> Choose a structure that minimises the need for repetition.

> **Don't say "as described above"***

Throughout your work, it is always better to use *forward* references rather than backward. Indicate when you first describe something that it will refer to multiple downstream applications. Readers prefer to keep moving forward through the paper rather than having to look back.

this is good general advice, it isn't specific to methods sections.

Next, write the rest of your experiments!

Pretty obvious really! Seriously though, once you have written the parts that are common to multiple elements of your story, the best order for the rest is the order the data appears in your figures. This is not a hard and fast rule if you can make the methods shorter or easier to digest by ordering it in a different way, then do that.

> **You do not need to mention the figure location in your methods.**

Statistics and data analysis

The data analysis and statistics sections of the methods are usually problem areas for new writers. Your methods section must describe how you got from the raw data to the final numbers used in your statistical tests. It is no use showing the reader a graph if they don't know where the numbers came from!

Include a clear description of the *type of your study*, e.g. "human observational study". Give a clear description of the variables measured, how you measured them, and what their names were. If appropriate, identify the variable defined as the primary outcome measure. Also, include details on how you analysed subjective data. For any digital images, include all the details about the software used as well as the steps in post-acquisition processing and analyses.

Provide a clear *description of the statistical methods* used. How you do this depends on the type of study, but here are some general tips for statistical methods:

Statistical methods details required

- What descriptive statistics were used (means or medians and SD. 95% CI, range or interquartile ranges).

- Provide the names of all the statistical analysis methods used.

- Describe how you checked the assumptions of your tests, e.g. testing for normality, and what methods you used when the assumptions were not satisfied.

- If you used an advanced statistical method, then provide the reference to the book or a paper that established that method.

- Provide the level at which you defined as the threshold for significance (e.g. "$p < 0.05$ were considered statistically significant).

- Clearly state what you defined as the experimental unit and describe any technical repeats performed. "Experiments were repeated in their entirety three times, with three technical repeats performed per experiment."

- Describe the approaches used to avoid bias. How did you randomise your sample acquisition and processing? At what stages were blinding employed?

- Include a statement on how you calculated your sample size.

Note that these points aren't my personal opinions, they are all based upon discipline-wide-rules.

Additional rules

In the following situations, there are also official checklists that need to be followed and some journals require these checklists to be uploaded along with your manuscript:

- Experimental Animal studies –   NC3R checklist.
- Observational studies –   STROBE checklist.
- Clinical trials –   CONSORT checklist.

Methods Figures and Tables

Tables

The general rule is, *if you can deliver table contents in four lines of prose or less, then you should write it rather than use a table.* This rule means you usually *don't* use a table for buffer or media formulations or PCR cycle conditions. In a University paper you have more flexibility, but I would still follow the same premise; unnecessary tables are disruptive to the flow of your document, reducing the enjoyment for your readers.

Big Tip

Don't go crazy with tables in your methods section.

If you do use a table, make sure you format it appropriately (generally very minimal). Also make sure to extract maximum value by including *all* the details, e.g. for PCR primers include not only the sequence but also the T_m, the amplicon size, the location within the gene and reference to the source of the primers. Appendices (for dissertations) or supplementary methods (for journal manuscripts) can be considered for things that aren't directly part of your story or for methods that relate only to supplementary figures. Examples where this might work, are cell line validation, extensive lists of buffers and product source lists. These details should always be included but they take up a lot of room and are likely only interesting to people attempting to replicate your study.

Figures

In general, you will not need figures in your methods section. You do not need to use figures for any standard setups or common experiments. However, if you have used a new or complex experimental approach then diagrams can help a reader appreciate how things were set up.

Depending on the type of manuscript you are writing, you might have one overall set up e.g. a new model system, that supports all the data, or you might have many standard approaches but with one or two newish techniques that contribute to parts of individual figures. How you deal with describing the set up reflects this difference. When you have the single model system scenario it may work in the methods. However,

Big Tip

Methods figures work best alongside the data in the results section.

my general preference is for methods diagrams to be placed as close to the results that came from the experiment as possible. The methods diagram becomes panel A of the figure and the results panels B, C, D etc. I find this option benefits the reader as they can see the set up and results in one go rather than having to turn back to methods section.

Remember that in a manuscript, you must describe your figures in order. For most journals, methods come before the results section, so if you refer to a figure in the methods, it will have to be Figure 1 in the manuscript, which may not be the best place for it. To allow you to reference the diagram wherever you want it to apprear, you can use text only to describe the experiment in the methods without mentioning the figure, then in the appropriate location in your results section add one or two overview sentences to provide the point where you can refer to the diagram (I'll come back to this in the results writing section in the next few pages).

Once again, do not refer to any data figures in your methods section.

What to check when editing

The most common issue with methods sections is that they can be torture to read! Although they are naturally dense with details, you must still write in proper sentences, using proper paragraph structure. A common sentence structure that pops up but sounds terrible is "and then...and then... and then..." Watch for these or similar sorts of meandering sentences and take a little time to turn them into something a bit more readable.

The other common issues are missing the source information of a reagent or its citation, inconsistent formatting of SI units, and structural issues that lead to repetition. You are likely to spot the latter scenario in your first read through and, often, the simplest answer is to consolidate any time you describe the same approach into a new subsection. For the missing details there is no quick fix, you just need to be thorough with your editing. Similarly, for SI units issues, but here you might be able to use search and replace. The best plan is to write consistently from the start (of course).

7.7 Results Sections

Describing your results effectively can maximise the impact of your findings and help to get your message across. The results eventually become the easiest part of your manuscript to write, actually easier than the methods, but, in my experience, it's an area that early-career scientists need help. The usual problems are either that there are incomplete descriptions of the finding, or that there is not enough contextual information to allow reader follow the progression of the study. Let's try to avoid these problems.

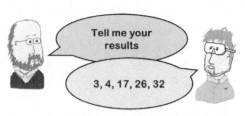

Reporting your data without any context means your reader needs to do more work

Surely, the results section it is just about writing down the results?

Yes, that is the basic premise and if you did just that it would be OK. However, that style is very dry and with just some small changes you can make your paper more enjoyable and accessible for the reader. Moreover, by writing your results in a way that guides your reader you can help them interpret the findings and thereby advance the narrative of the paper.

Hopefully, you have read lots of papers by this point but you may not have noticed how those papers differed in the way their results sections are constructed. Look carefully at the papers where you enjoyed reading the results and you will see some consistencies in the tone and rhythm of the writing.

Results Subsections?

Structure first – identify figure order and subsection breaks

Hopefully, at this stage you have made your data figures and have thought about the connections; how the story will *flow* from one figure to the next. As you write, you might find that your plan is harder to follow than you hoped, that the connections don't work. This can be a sign that should adjust the order rather than continuing struggle to connect the narrative. The connections from one figure to the next should feel natural and should be easy to describe.

If you write your results as one long list of numbers without any surrounding text, it will be hard to absorb. Therefore, the next step is to break up the section into accessible pieces (subsections). The length of each subsection varies depending on the data and its contribution to the story. Each subsection should deliver one key point fully but shouldn't be too long. Of course, too small a subsection can be a problem too. If you break up your text into tiny pieces, it will be more disruptive to the flow than it is

> **Big Tip**
>
> Print your figures or have them open on the screen as you write your results.

beneficial. Therefore, you are looking for a happy medium with around three paragraphs on average, roughly 350-500 words, and this usually means describing one multi-panel figure or perhaps two less intricate figures.

As always, a caveat: not all journals allow your results section to be broken up into discrete units, with some requiring one continuous section. However, whether there are physical separations between subsections does not change the content you deliver very much, only that you don't have the benefit of sub-heading.

Components of a subsection

The critical part of **every** subsection is the text description of the results. The description of the data you collected. However, each subsection could actually contain up to four other elements whose job it is to provide context, where necessary, for the results and thereby help your readers to absorb the information.

It will be extremely rare that you need all of these components in a single subsection. Which of the additional parts to include depends on the complexity of the data, the preferred style of your supervisors, and what is standard for your field. It is a balancing act where you are trying to provide enough to help the reader follow the story but not so much that you bore them by over explaining obvious things.

Subsection Elements
Subheading
Why?
How?
What (text)?
What (data figures)?
Wrap?

Element	Usage	Purpose
Subsection title	**Always***	**Signpost, advance the story**
Why clause	Infrequent	Establish the rationale for a change in direction
How clause	Usually	A very short overview of methods to make the data easier to understand
Results	**Always**	**Text description of findings including descriptive and inferential statistics**
Figures and tables	**Always****	**Evidence to support your conclusions**
Wrap / conclusion	Rarely	Concluding sentence to help clarify complicated or explain non-intuitive findings

*If allowed! If a journal requires one continuous section of prose without subheadings, then I recommend using "why" and "conclusion" sentences more frequently to help frame the narrative.

**In old papers you might see "data not shown", that is not allowed in modern science

Adjust subsection length to increase impact

Although each part of your study will contribute to the overall conclusions of the manuscript, their relative contribution is rarely equal. One easy way to draw more attention to the important points is to change the amount of text associated with different parts. Although the data part of the subsection will not change, you can use more of the framing to deliberately increase emphasis on the most important part. The first description of a new and striking finding might get a full subsection with all elements, whereas the triangulation-type experiment in a different system or using a different approach might only need the results and figures part. Considering the relative contribution to the conclusion becomes particularly important when there is a small experiment that you *could* describe in a single line but which you don't want the reader to miss. Adding some extra context or interpretation to bulk it up can help the overal impact.

Subheadings

Most journals allow use of subheadings to break up your prose. The subheadings are there primarily to help your reader find information, but you can also use them to help advance your narrative. Based on these different functions, you have three options depending on what you want/need to achieve.

Types of subheading

1. Simple signpost (for very simple papers)
2. Objective signpost (to establish a model system or validate methods)
3. Results statement (for reporting data)

Simple signpost

The simple signpost helps the reader find the subsection they are looking for but nothing else. They can be as simple as the name of the technique used to generate a specific set of the findings. Example simple signposts: "RT-PCR data", "Tensile testing". You will see this heading most commonly in bioengineering or biophysics type journals.

Personally, I don't like them. These sorts of minimalist subheadings do not add much value beyond breaking up the text. I think they are a wasted opportunity as it takes only the smallest of effort to convert them to something that contributes more to the narrative and reader experience. I recommend only considering simple signposts when your story is linear with all the parts asking the same question without adding additional

Simple signposts
Guide the reader to the data
but doesn't advance the story

nuance.

Simple signposts don't tell the reader anything about the objective of your experiment or what the data mean, use *why* and/or *wrap* sentences (see below) more frequently.

Objective signposts

Rather than merely highlighting the technique, an objective signpost lets the reader know *why* you did a set of experiments. Example objective signposts: "Analysis of changes of Per2 mRNA abundance in response to serum shock", "Determination of changes to material properties after plasma treatment." Hopefully, you can see how these examples would be more helpful for a reader than just "RT-PCR" or "tensile testing"; they let the reader immediately move on to finding the answer to the question. Objective signposts are widely used in engineering and medical papers, and also in any biological papers where the subsection contains data that is descriptive, involves direct measurements rather than requiring interpretation.

Even if overall you prefer the *results statement* types of subheading (next), objective signposts will be what you need to describe a new experimental tool or to describe data that establish a model system; i.e. where the experiment is not testing a hypothesis but is necessary to set up the next part. You also see effective use of objective signposts when the authors are describing multiple tests on a single thing, where the core hypothesis does not change direction as you progress through the paper.

As objective signposts sub-headings do not say the conclusion from your experiment, the reader needs to get that information from the rest of the section. This means you will be more likely to need a *wrap* sentence in subsections where you have used an objective signpost.

Result statement

In terms of advancing your story and helping emphasise your findings, subheadings that contain the *answer* to the question your experiment asked are the most effective. I class these as *results statements* and you can think of them as essentially the same as you would a title of whole paper. Using the same examples as the previous two options, we might have subtitles of "Per2 mRNA abundance increases in response to serum shock" or "Plasma treatment does not influence tensile strength of material X".

Declarative statements can feel too bold when you first start writing; however, subheadings are an opportunity for you to show the reader what your data mean before they reach the discussion. If you are writing a student paper or thesis, this is a chance to let your examiner know that you understand your study. Indeed, if I am marking a piece and a student has chosen to use an objective signpost when a result statement would have been better, then I worry that they haven't thought about or might not understand what their results mean, or that they aren't confident in their findings.

Use subheadings to state your findings and advance the narrative

In case you haven't guessed, the results statement is my preferred style and is what you can expect in most types of biological science-type studies. Importantly, your work doesn't have to be complex for you to use this approach; it works just as well in simple studies.

One extra point to note. If your experiment provides indirect measurements, your description of the data should be limited to what you actually measured. However, to advance the narrative, you will make an inference about what those data mean. In the results, you can do that in two ways; in the subheading or in the wrap. Don't forget that you will need to cover the limitations of these interpretations (likely in the discussion).

It's all about the mix

While declarative statements are my preferred option, a manuscript will almost always contain a combination of the different styles. Objective signposts are used for tools development figures often early in the manuscript, while results statements used later in the paper for reporting and interpreting the data obtained.

> **Big Tip**
>
> Don't try and force it. Use the sub-heading style that best suits each of your sub-sections.

"Why" (or Link) Sentences

An example *why* sentence might be as simple as: "Next we tested the hypothesis that…", this sort of phrasing gives you a chance to be explicit about the hypothesis, objective or aim of the next series of experiments. The purpose of the *why* sentence is simply to help the reader follow the progression of the story. However, you don't always need them and you should include one if it is beneficial.

After you decide on a subheading you can decide what other parts you need. For each subsection, you should think about where your readers will be in terms of the narrative. How obvious is the purpose behind the next subsection? If you have used a objective signpost, it might be very clear but if you chosen a results signpost then you might a little bit of shaping to help the reader follow your objective.

The single sentence or even just a clause is the most common type of *why* sentence and is usually all you need. The bare minimum to carry the reader into the data part of the subsection. Occasionally, you can go deeper, connecting the next set of data either to previous studies or to earlier data in your manuscript; "These data raised the possibility that… and previously it has been demonstrated that … (ref), therefore to test this we…."

Use the longer format

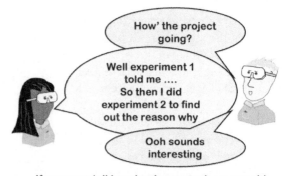

If you were talking about your work, you would connect the different parts together to help your listeners understand the progression. Your writing should do the same.

sparingly, you should not make the results longer than they need to be. Indeed, some authors never use why sentences, so do not be surprised if your supervisor removes them from your draft. You likely won't need a *why* sentence for your first couple of subsections as you will have established your main aims and hypotheses in your introduction. Only include a *why* sentence when they serve a useful purpose, where it helps the story.

Example "Why" sentences

Simple version
"Our next goal was to assess the integrin matrix receptor expression profile of BEP2D and NHBE cells..."

Connected to the literature version
"Preliminary visual analyses suggested that the eyelid regions were missed with a higher frequency compared with the rest of the fact. As the eyelid are particularly prone to skin cancer development (refs), we..."

Change in direction: connect to past literature and to your data*
"Human keratinocytes migrate on a LM332-rich matrix (refs). In contrast, the fibronectin in the matrix of mouse keratinocytes appears to inhibit LM332-mediated cell motility (refs). The presence of both fibronectin and LM332 in the matrix deposited by alveolar and bronchial epithelial cells led us to next compare the migration behaviour of both cell types on a variety of substrates."

*The final version is quite long; use something like this very sparingly.

"How" sentences

The *how* sentence or clause tells the reader, the approach used to generate the data. This should be in the form of a simplified, high level overview. The *how* sentence serves as a little flag to the reader to help them identify which section of the methods they should refer to if they want more details. *How* sentences are much more common than *why*. I use them in almost every results subsection*.

At this point, you might be thinking that including a *how* sentence in your results will make your writing repetitive. This is a danger to be aware of, you must keep any *how* sentences as short as possible. Before you dismiss them entirely, consider your readers' needs. The results are the reason that many readers are interested in your work, and, therefore, your readers will often read the results sections before they read the methods. Adding a brief overview of the methods in the results can help the reader follow the flow of the experiments. The *how* sentence can also be used to remove ambiguity when multiple approaches ask different parts of the same question. In simple terms, the *how* sentences will mean that your readers can get straight into the action, absorb the information more rapidly and generally enjoy reading your paper more. Everything about paper writing is about keeping the reader happy!

If you are writing a thesis or dissertation, you should also think about your examiners. In longer-format writing, your methods and your results are often in separate chapter and quite far apart. There are also likely to be many more methods used throughout the body of work. A short *how* sentence will stop your examiner from having to flip back through 50 pages to find the relevant section of your methods.

If you feel that the description of your data would benefit from a diagram of the experimental setup, then the *how* sentence is a way for you to include the figure reference besides the results rather than in the methods section. This is usually the most effective and easiest for the reader to interpret.

The rare occasions when I don't use one are when the subsection title contains all the information requires, or if using the same technique throughout the manuscript, or in very short format papers using a very limited repertoire of approaches.

Example "How" sentences

Single clause

... Fluorescence-activated cell sorting analyses indicated...

Full-sentence

... To do so, we plated BEP2D cells on the matrix deposited by iHEK cells supplemented with fibronectin at 1, 2, or 5 µg/ml...

Referencing a diagram

... We next wanted to determine what happens to LaNt α31 expression and distribution in a more physiologically-relevant system, a porcine ex vivo three-dimensional (3D) alkali wound model (Fig. 4A) ...

Data description

The data description part is where you tell the reader all the results in clear, definitive terms. The challenge in results writing is that a dry statement of just the numbers is hard to read and even harder to absorb. Therefore, you should aim to lead the reader through the data in a way that is easy to read, easy to follow. Your goal is that no sentence should need to be read more than once. You guessed it; you should wrap the data into a narrative.

The sentences themselves should focus on the key observation. . Tell the reader the *direction* and *magnitude* of the differences (X was three times larger than Y) and provide the *actual numbers* (descriptive statistics) and *confidence levels* (inferential statistics). The stats results numbers are best delivered in brackets at the end of the sentence (see example box). This allows the reader can focus on the story of the data. Make sure you use the appropriate number of significant figures to reflect measurement error. Note that you do not need to repeat any of the data presented in a table.

The results text should be able to stand alone and completely describe the findings, a reader shouldn't have to refer to the figures to understand the whole

Big Tip

Refer to every figure panel in order. Usually at the end of the sentence describing the data (Figure 1A).

story. The data figures contain the *evidence* to support your conclusions. I know it sounds odd to imagine your readers ignoring your lovingly crafted figures but appreciating that the figures aren't strictly necessary, means you change the way you write sentences.

Instead of saying "As shown in Figure 1C ...", it is usually better to write your results as "the results revealed... (Figure 1C)".

Compare those options, and you will see that the first version emphasises the data *location,* whereas the second places much more emphasis on the interesting stuff, the *results.*

Focus your sentences on what the data mean.
Use clear, understandable statements to deliver the *direction* and *magnitude* of the effect, then report the descriptive stats, results from any stats tests, and provide the figure location in parenthesis.

Remember you must to reference and describe *every single figure panel* including supplemental figures. The rules are that you must reference figure panels in order they appear; figure 1B has to be referred before 1C. If this means you need to change the figure to suit the flow of the story, then do not be afraid to do that.

One of the traps that new writers can fall into is that the focus on achieving P values below a target threshold means that they forget that P values are measures of confidence and not absolute measures of importance. My advice is to organise your sentences so that they focus upon the *biological* or other real-world importance of the data rather than emphasising the statistical significance. A related problem is writing two sentences; one about the differences or correlations and then another about the statistics. You don't need to do this. One sentence is all you need.

Example data description sentences

...limbal-derived epithelial cells expressing LaNt $\alpha 31$ GFP 2D area was approximately twice that of GFP- expressing cells and non-transduced pCEC (2D area + LaNt $\alpha 31$ 2720 ± 720 μm^2, pCEC 1230 ± 380 μm^2 and +GFP 1280 ± 280 μm^2, $P < 0.05$, Figures 5A, 5B).

We next assessed single-cell motility by plating the transduced cells at low density on uncoated dishes and then tracking motility over 2 hours (Fig. 5C). These analyses revealed the LaNt $\alpha 31$ GFP-expressing cells display approximately 50% reduced cell migration rates compared with controls (+LaNt $\alpha 31$ 0.46 ± 0.14 $\mu m/min$ versus

> pCEC 0.91 ± 0.14 μm/min and +GFP 0.77 ± 0.04 μm/min, mean
> values for all donors, $P < 0.05$, Figure 5D).
>
> *Note: the text is assembled to focus on the story rather than the*
> *details, but also that the actual data, SD, P values and figure*
> *locations are all provided in parenthesis.*

Data Figures

Which figures go in the main text, which go in the supplemental figures?

We touched a little on this in the figure preparation section of the book, but I have come back to it here as it is when you are assembling the manuscript that you are in a better position to decide whether a figure panel is integral to the story or if it should be included only as a supplemental figure. This decision will not dramatically affect how you write about the data in the text of your results section except that you might choose to write slightly less about supplemental figures and more about core figures as a strategic mechanism to reflect their relative importance to the narrative.

Big Tip

Most journals now allow unlimited supplemental figures. There is no reason for "data not shown".

Supplemental figures will get read and reviewed, the decision about which elements go where is just about impact. The temptation is to include everything as main body figures. That's fine if you have unlimited room such as in a thesis or when have relatively small amounts of data that you want to deliver. However, most journals have restrictions on the number of figures you are allowed, and sometimes the impact of a figure can be improved by removing some non-essential elements and, in so doing, place greater emphasis on the parts that matter most. Using supplemental figures well can help to deliver your message.

Some situations where it might be better to move a figure to the supplements:

- **Reagent validation**. Things like antibodies specificity tests, primer melt and efficiency curves, cell line validation, or other tests of your experimental system need to be included somewhere in your manuscript. However, these data probably don't directly contribute to the narrative and so can be moved to the supplemental figures if desired. This is especially true if they are not "new" reagents produced but rather validation of things produced and described in detail elsewhere.

- **Repetition of an experiment in a parallel population**. If you are reporting on findings in one cell type, you might use the supplemental figures to contain the data from the same experiment in another similar cell types. These extra data are valuable for validation, triangulation and generalisation of your findings, but they might not add anything new to the central story.

- **Negative data**. Negative data, findings that show something *doesn't* cause an outcome, are just as important as positive data and they should always be reported. In situations where you have tested many outcome measures or treatments and some have no effect while others reveal interesting findings, then you will want to focus your story upon the interesting parts. All the negative or "no effect" data should still be included, and you should write about them, but likely moving them to supplemental figures will help focus and streamline the story.

Embedding Figures into the document (often not necessary)

When you upload a manuscript to a journal the figure files are usually uploaded separately as .tiff, .pdf or .eps type files. This means that you *don't need* to embed the figures in your text. Keep them as separate files throughout the writing and editing process.

In contrast, if you are writing a student manuscript, including a dissertation or thesis, you probably have some options. If you are using Microsoft Word or Apple Pages programs, you could choose to embed your figure in with the text. If you do that, be careful of the following points:

- Make sure that you are not compromising quality. Check that in the final output you haven't dropped the resolution of your images and that images and labels present properly.
- Make sure you are following the figure rules. Be especially careful if you need to resize your figures as things like the font and line sizes may have changed.
- Don't annoy your co-authors/supervisor. When you embed figures, use page breaks and anchors to make sure that the figure stays in the appropriate place within the page. Figures jumping around the page as you are editing is annoying!

My advice: make figures independently of your text with each figure on a separate page (with the figure legend if desired, although these usually go at the end of the manuscript text). Only combine the figures with the text at the very end of the writing process.

Big Tip

Don't embed your figures unless your instructions demand it!

Wrap Sentences

"These data demonstrate…" or equivalent summation sentences can be included at the end of a result subsection to round off the paragraph. Including one can be an opportunity to hammer home your interpretation of the data or to add emphasis to specific observations. Occasional use of wrap sentences can help to reinforce a specific part of your message or can be used to clarify a complex set of disparate data. If you use a results statement style of subheading, then it is not usually necessary to repeat your conclusion in prose form.

Putting Results Sections Together

You won't not need to use *every* component in *every* results subsection. Only use what you need to advance your story. Below is an example of how the different parts could come together to tell a complete story.

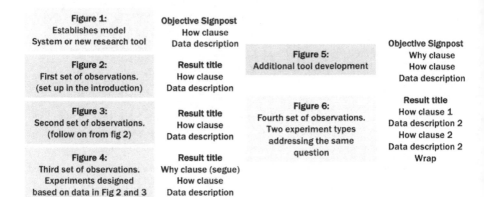

What to check when editing

Simple things first: make sure you have *written* about every panel of every figure and have presented them in appearance order. Make sure you have included the magnitude of the effects, summary statistics and P values for your different populations.

Flow

Once you have checked that everything has been written about, it is time to edit at the sub-section level, checking that each part is as clear and succinct as possible. The biggest thing to look for is the progression through the story, does it transition logically from beginning to end? Will the reader be able to understand the reason why you did the experiments and what they mean? Getting this overall flow to work well is more important than any individual sentence being perfect.

Significance

In my experience, many new writers tend to write, "statistically significant" (or not) in every sentence. However, this makes for harder to absorb sentences. Importantly, you should report the P values therefore by also stating "significant" you are effectively repeating yourself. At the editing stage, I recommend looking at every time you have used the word "significant" in the text and carefully analyse whether it makes the sentence better or not. Ask yourself is there a reason to emphasise the *confidence* in your findings rather than allowing the *P* value to tell the reader that part of the story? Using significant very occasionally for impact, to highlight some specific difference or lack thereof, can be very good, but if you find that are saying "significant" in every sentence then you are in a situation where editing at least some of those sentences will improve your work. Always make sure the primary focus of the results sentence is on the biological or real-world meaning of the results!

7.8 Discussions

Discussions tend to vary more dramatically between different papers and, on first impression, it is more difficult to identify a single formula that works for all. With this flexibility, it can be the hardest section to get going. Therefore, in this section, we'll discuss some general concepts then establish a framework to use as a starting point. With the broad structure in place, you can either build upon or reduce sections to suit your project and message. Hopefully this will make the writing your first draft easier.

What is the purpose of a discussion?

The discussion's role in a paper is simply to give readers guidance about what was accomplished in the study, the scientific meaning. It should define how the new findings fit in context with what has gone before so that readers can appreciate how the work advances the

Big Tip
The discussion is not a second results section.
Discuss don't repeat.

field, what the new results mean for the bigger picture. It should set the limitations of what can and cannot be interpreted. It could also suggest what should come next.

Remember that you (should) know more about your subject than many of your readers, including knowing more of the relevant literature. The discussion is a chance for you to use your knowledge to add value to the reader experience through helping them interpret the findings in a way they would not otherwise have done.

Discussion Do's and Don'ts

Do

Describe the value of your work (without exaggerating).

Stay focused on the core message of your manuscript.

State if your conclusions support your hypotheses.

Consider other ways to interpret your data.

Attempt to explain unexpected results.

Present potential reasons for differences compared with published work.

Take a holistic view as far as possible.

Don't

Don't repeat your results.

Don't refer to data figures.

Don't discuss each experiment as an independent result.

Don't go on too long; stay focused on the message!

Don't make it too short: address all the things you need to.

Don't overreach in your claims.

Don't introduce new terms or deep ideas.

Don't focus too much on limitations.

Don't ignore any controversial findings.

Discussion Structure

The discussion is the most flexible section of a manuscript, its length, structure and flow should be driven by how complex your story is and what your findings mean in relation to the wider literature. Surprising or controversial findings will need more discussion than those that were exactly what you expected. More linear studies that achieved a narrow and well-defined objective may not need much discussion at all.

Now, with those caveats out of the way, we can look at what I consider my *default first draft* structure. It works for most studies and is a good starting place.

Synopsis	~130 words*	Short recap of key findings (if necessary)
Biggest Finding	~190-300 words	One or two paragraphs addressing the most important aspects of your work in relation to the wider published literature. This should be your largest section
Other important findings	~150 words each	Shorter sections on any secondary findings (if necessary)
Limitations	~150 words	Acknowledge anything that restricts your ability to draw robust conclusions, and/or address alternative interpretations
Wider implications	~170 words	Describe what your work means in a broader sense
Conclusions	~100 words	Short, focused ending that captures the key message of the whole paper

*Word counts are indicative only!

Paragraph 1: Synopsis

Summarise the goal of the study and how your results* addressed that goal. Don't repeat your results here, rather aim to combine your key findings into a simple message. It should be short. A two or three sentence paragraph should do it. **You should not refer to data figures or specific numbers** here, or indeed elsewhere in your discussion, the synopsis is about themes and generalisations.

> **Big Tip**
> If your synopsis is over 150 words, it probably is too long.

Connect your synopsis with the rest of the discussion by using phrasing that emphasises the points that you will delve deeper into the following paragraphs.

If your study is quite short (1-2 figures) or simple, you might not need a synopsis at all. In contrast, if you are delivering seven figures plus supplementals, with five plus panels per figure, then this first paragraph is really important in bringing all your results together into a single message, reminding the reader how the pieces of the story build on each other and setting up your big-hitting paragraphs to follow.

Finish this paragraph with a forward-looking wrap sentence to lead your readers into the next point. Now, let's move onto delivering the big stuff (see what I did there?).

Data, results and findings are often used synonymously but really data refers the measurements before analysis whereas results are the analysed data and findings are the interpreted results. In a discussion you shouldn't need to write about data except justifying outlier decisions, mostly you will discuss results and make conclusions based on findings.

Who owns the results?

Which of the following options is correct?

"Our results show that ..."

"These results show that ... "

The answer is the second one. At this point in the paper the results have been discussed and are now owned by the *scientific community*. They aren't yours. You can't own results. Therefore, you should use passive voice to describe them.

You can own thoughts. Therefore the following would be appropriate;

"Based on these data, we hypothesize that..."

Paragraphs 2 (& 3): Discuss your biggest finding
Paragraph 4: Discuss secondary findings

Next, you bring out the big guns; what have you discovered and how do those discoveries fit with the literature? Assemble these points in order of importance; the most biologically meaningful or the most impactful, exciting, or surprising finding first followed by any secondary, but still interesting, findings.

Don't direct your reader back to the data figures! Advance the story onto discussing what the data means!

This should mean that the second paragraph of your discussion directly discusses the point you made in the title of your manuscript and provides the clear statement supporting or refuting the hypothesis you stated in the introduction. This connectedness is part of the reason why it is best to write the intro and discussion alongside each other.

Aim to devote space in proportion to relative importance. The main finding should get more space than any secondary findings. Moreover, if your paper is quite simple, then don't add extra paragraphs here for no reason, discuss your major finding and move on. Follow the usual rules: one key point per paragraph, fully supported with a clear topic sentence and a wrap sentence to connect to the next paragraph.

Big Tip
You don't necessarily have to discuss everything! Focus on the most import and impactful findings.

When you are planning what to write,

focus your early energies on identifying the key messages you want to deliver, then write accessible and, ideally, relatively short sentences to use as your lead-in for each paragraph. After your short, punchy topic sentence, stay on the same message for the rest of the paragraph. Doing this will make sure that you focus your work on what you want your reader to take away rather than going off on a tangent.

Content

The meat of the "big finding" paragraphs is where you deal with whether your data supports or refutes your hypotheses *and why*. Take a holistic approach and consider the whole of the results together, rather than focusing on individual experiments in this part. Before you attempt to write your own version, do some focused reading of good quality papers to identify different ways authors approach this part.

Use these second and third paragraphs to highlight the advance. Structure them to maximise the emphasis on why your findings are *important* or *useful*. State how have they changed or advanced what we as the scientific community think about the topic. The biggest killer to a paper, which will get it rejected from any decent journal, is if the work doesn't add anything of value. Remember here that novelty isn't the same as value, you need to emphasise why the new findings are important. Be careful not to go too far. Your study might be an improvement that adds mechanism or robust validation in a more physiologically relevant system (validation is valuable), but don't try and sell it as bigger step forward than it is.

In these big findings paragraphs it is essential that you cover your competitors' work. One thing for sure; your reviewer won't appreciate it if you have ignored their seminal work in the field. Wherever your data fits with or disagrees with other people's findings you should cite those works and discuss why any differences may have occurred. What have you added/changed? Why is your study better? Has there been anything published in between, which, when combined with your results, leads you to a different interpretation? A relatively simple way to discuss your findings is to draw parallels with studies of related proteins or drugs, or to similar studies in other organisms,

Be careful with assertions of novelty! Your reviewers should know the literature and will spot where you have exaggerated!

tissue types or diseases. These are all things that are worth stating explicitly to make them clear to the reader who may not be aware of them.

The discussion is also for putting the findings into a wider context. One way you might want to consider doing this is by updating a model or diagram to capture how all the pieces of the puzzle come together. Make sure the diagram actually does help though, don't just put one in for the sake of it and be careful that the diagram is not be too much of a stretch relative to what you have discovered. If you do include a diagram, don't forget to reference it in the text and remember that the diagram should illustrate rather than replace the text description.

And finally, don't forget your citations! These core discussion paragraphs will need to be supported by *lots* of primary data references. Every token sentence within the body of each paragraph should be talking about your work in relation to others. If you find that don't need to cite many other papers then that it is a warning sign that you are either repeating your results or taking too narrow a view of the findings!

Paragraph 5: Discuss limitations and caveats

No experimental system is perfect, and this restricts how strongly or definitively you can state your conclusions. You need to clearly define these limits and explain why you are prepared to accept them in your discussion. I recommend approaching this in one of two ways; either use one dedicated paragraph to cover limitations, or combine the description of any limitations with the discussion of the key findings.

So that you don't sound too negative, aim for a balanced approach, one which emphasises the strengths of your study while simultaneously acknowledging the areas where you cannot be as confident. To do this, phrase your sentences so that they focus on what you *can* interpret. This balance comment also extends to how much space you devote to limitations. If your limitations sections extend for longer than your major findings, then you will leave the reader with the impression that your experiments weren't very well designed! There's not much chance of convincing an editor to publish your work if *you* aren't convinced the findings are strong enough to be believable!

What I like to do here is describe the different ways the data could be interpreted. Rather than present these alternative interpretations as a problem, I would instead detail how to differentiate between those alternative interpretations in the future and present arguments as to why you are willing to accept these limitations at this time. For example, something like; "To resolve this dilemma would require 100 extra human samples, which is unethical to request at this time."

Whenever you describe limitations that you could have addressed easily, my suggestion is to do the experiment rather than writing that you should have done it! Every paper needs to end somewhere, but leaving small things unfinished makes the story weaker and reduces its value.

A quick note on student assignment papers

I frequently encounter in student papers limitations section are overly long, overly negative and, importantly, fail to present solutions to the problems. One way to cut how much you write is reduce the commentary on superficial things. Avoid obvious comments about time limitations and the impact that has had on the number of biological repeats (markers know there was a time limit). Your markers will certainly know that the data might be more reliable if gathered by experienced researchers (I've seen this included as a limitation a lot more often than you would expect). Using more than one sentence to say any of these self-explanatory points won't add any *valuable* insight. If you do feel it is necessary to include something about sample size, then do so in a way that is short and focused, "sample size limitations preclude robust statistical inferences". Better still, indicate what sample size you would need to make future data reliable "based on the data spread and effect size, 10 further samples are required to achieve 90% power". If I am marking your work, I want you to show me that you have thought about the problem *and have a solution*.

Paragraph 6: Discuss wider implications

As you progress through your discussion, you should steadily get wider with the implications of your findings (i.e. the mirror image, to your introduction, which goes from wide to narrow). Put your findings into their real-world context and demonstrate the *big picture* value. For biological research, this part can often include the wider implications for clinical needs; for example, you could describe how the work you have described will open an avenue toward therapeutic development or how it should change clinical practice

Be positive about your findings but don't overstretch their importance. Not acknowledging any limitations will make your interpretation seem naïve.

The wider implications part of your discussion is also a place where you can talk about the next big question(s). A specific agenda for future research based on the questions generated by the current findings is much more helpful to a reader than vague suggestions. Reviewers and readers appreciate how specific research recommendations will advance the field. An alternative approach to talking about next steps is to integrate these types of phrases into the limitations paragraph. For example, you might stress the future value of making an animal model as a counter a current limitation of the absence of immune cells in my cell culture model.

As with limitations, don't go too far here. A long list of what needs to be done next could leave a reader thinking "what have they actually achieved?" e.

Paragraph 7: Conclusion

Round it off. Tie it up with a neat little bow. In positive language, describe your core finding and what it means in relation to the big picture. This **short** paragraph should be a slightly expanded version of the last line of your abstract and should connect to one or more of the motivations behind your research, the ones you established in the first paragraph of your introduction.

You might be tired and excited to finish by this point in your writing, but make sure your conclusion is as perfect as possible, it is the last chance to influence your readers' and reviewers' opinions and will be the final thing you need. Indeed, some journals view the conclusion as being so important that it gets a separate section with a subsection header.

One little point to note here, you shouldn't need to say "in conclusion". If you feel that the paragraph needs this phrase, it is likely to be a sign that you aren't being conclusive enough!

Final comments on Discussions

Or should that be conclusion? A discussion in a manuscript or project should be the length it needs to be. I've seen three paragraphs work well and equally a twelve-paragraph version that didn't feel too long, and those were both by the same author on similar topics. Don't force it. Use the style that works for the story that you are telling. If something needs to be discussed, then discuss it. If things stand-alone, are complete or are obvious, then let them lie.

What to check when editing

My biggest tip for discussions, as with other sections, is to set out a framework of topic sentences and then get a draft down without worrying too much about writing the perfect sentence. Once you have something in place, it is easier to edit, edit, edit until the message is clear and compelling. Some things to check:

Have you addressed the hypothesis?

If you set out to test a hypothesis, then your discussion should come to a conclusion about the outcome from your tests. Either the data supported or refuted your hypothesis. This should be written in a clear and understandable way.

Avoid discussing single experiments on their own, in order

Your discussion should be looking at the cumulative knowledge throughout. When you are editing, check that any time you mention a specific experiment you are doing so in a way that highlights its contribution to the combined story or are dealing with a specific limitation. Remove all references to primary data figures.

Make sure it's not too narrow

I find that new writers first draft often spend too much time focused only upon *their* findings, failing to connect those new data with the wider world. The only place you can be self-centred is in the synopsis and conclusion. During the editing stage, make sure you have added enough connections between your work and the literature that everything is connected.

Check that you aren't being overly speculative

A little speculation is good, but you can go too far. When you are editing, look at the balance of what you have written and try to avoid over-interpreting what you have found. Avoid including things just to show off your knowledge; every point should add value.

Watch that you haven't identified too many unknowns

If more than half of your discussion is saying "we don't know..." in a variety of different ways, then the overall feel to the reader will be that the study is incomplete. It isn't surprising to have unknowns, good research often raises more questions than answers, but do you really need to discuss them all? My advice in editing is to reduce the unknowns into a single paragraph and set yourself a rule that whenever you say "we don't know" you also include some statement of how you would address the gap in the knowledge.

Check that you haven't added new information

Make sure that your introduction and discussion are connected. You shouldn't be bringing in completely new material into the discussion. It might be appropriate to fix this issue by editing the introduction rather than the discussion.

Check agreement between title, abstract, aim and conclusion

This final editing point is really important. Your manuscript should be telling a consistent story from start to finish. Your concluding paragraph should have different phrasing from the abstract and title but should essentially be saying the same thing. The last paragraph of your introduction where you state the problem, aim, hypothesis or objectives should also agree with the final conclusion. During editing, it is easy to focus on one part of the text in isolation, but if you don't get these four points to connect then the whole feel of the paper will be disjointed.

Chapter 8: Editing

8.1 Document-level editing
Structure and content editing to focus the story and improve flow.
8.2 Paragraph-level editing
Editing to improve content delivery at the paragraph level.
8.3 Sentence-level editing
Making individual sentences more effective and impactful.
8.4 Word choice
Maximising impact by improving word selection.
8.5 Final checks
One more run through before you hit submit.

8.1 Editing Stage I: Document-level

With the first draft complete, you are an important step closer to submission. Now it is time to start making your work even better! The next few pages hold some general tips for things that will make your writing more impactful and enjoyable to read.

The following tips are applicable to all forms of writing both inside and outside science.

> **You are aiming for writing that:**
> Conveys your message unambiguously.
> Does not require anything to be read twice.
> Sounds good, with pleasing rhythm and flow.

General Editing Advice

> **Big Tip**
> Keep an early version of your document in case you change your mind about deleting anything.

Get the story right, and the rest will follow. Editing is a stepwise process; you start broad on the structural elements and move gradually toward smaller edits such as improving word choice and removing typographical errors. We edit in this order, as it is not unusual to decide to move things around, to delete whole sentences or paragraphs and to add entirely new material. These large changes mean that it can be a waste of time editing at the sentence or word level, if you later to decide that you don't need those sentences anyway.

The longer the piece of work, the greater the scope there is for making major changes to improve impact and effectiveness. Content editing in a doctoral thesis or dissertation is very likely to involve making both substantive and substantial changes. Do this major editing with input from your supervisory team.

Take a break between completing the draft and beginning the editing process

When it comes to editing, you need "fresh eyes" to spot mistakes. Although you know what you wanted to say, it is very common that what was in your head and what actually ended up on the page doesn't quite match.

Ideally, you should aim to

Taking a break between writing your draft and beginning editing will help you to be more objective about what works vs needs amended.

have about one week between completing the draft and coming back to edit. However, we need to be realistic; you are probably working toward some deadline or have pressure from your supervisor. Assuming that is the case, aim to complete a draft at the end of the day then sleep on it before coming back to edit. If you are writing a long-format document, like a doctoral thesis, then it is most effective to complete the draft of one chapter then work on a different part of the thesis for a few days before coming back to edit.

Read some papers to get inspiration and to practice being critical

Editing and writing are different processes, you are shifting from thinking about content and instead thinking about delivery. Editing skills are something you continue to develop throughout your studies and can also be honed at journal clubs or in writing groups. If you are able to build in a time gap between completing the draft and beginning the editing process then it can be useful to catch up with the literature and transition into *editing mode*. Think critically about any ways you could improve the work in terms of the way the message has been framed. You will find that it is much easier (at first) to be critical of others writing compared with your own. The more you do it, the better you will become at improving your own work.

You don't always have to criticise, if something is written well or is particularly effective, then try to identify what made it is that writing *work* and adopt that technique yourself.

Read the whole piece without stopping to edit.

Once you are in editing mode, the first thing to do is read the whole piece from start to finish without correcting any of the minutiae.

If you have been working entirely on a computer screen up to this point, it can help to print out a hard copy and then do your edits offline. Indeed, I find it helpful to perform this first read through in a different location entirely; away from my office in a comfortable seat somewhere.

This first read is all about getting a feel for the entire document, to identify more extensive issues that need work. Indeed, the goal is to read it in one quick go, so skim read using just the topic and wrap sentences of paragraphs. Resisting the temptation to edit every sentence as

Although you aren't writing a novel, you still need to have a clear "storyline" with a solid framework and natural progressions.

you go can be hard, so a compromise is to quickly mark where corrections are needed so that you don't forget to come back later. At this point though, your primary focus should be the big stuff (see box below).

Big Stuff!
Problems to identify in the first read-through

Irrelevance:

Remove anything that does not contribute to the conclusions of the work. Often, most of the first round of edits will involve deleting things that you thought you might need, but it turns out you don't. It can be hard letting go, but nobody wants to read something bloated and overly long.

Incomplete information:

You know the material really well, but do your readers? There is a real danger that you will assume too much prior knowledge or will forget to elaborate on a point. In your first read-through, try to identify any passages where the setup is incomplete or where more details are required for the deeper or more complex later parts to make sense.

Appearance order:

Look for issues in how the story builds. You are aiming to start broad and become more specific as you progress, then working back towards wide as you reach the conclusions. When reporting data, consider the thought progression between each experiment. Identify anything that isn't connecting well and then test out whether a different order could help.

Repetition:

There actually should be some repetition in your writing. Points raised in your abstract or introduction will be expanded in the body of the paper and briefly covered again in your synopsis or conclusion. However, there are two important points. Firstly, do not repeat the same phrasing anywhere (no copy and pasting sentences). Secondly, you should not have to make the same point at the same level of detail twie. If you haven't gone deeper in the subsequent mentions then there likely is a structural problem somewhere. Identify any time where you have done this and decide whether you need to go deeper at the second mention or whether you should delete the superfluos part. You don't need to repeat.

Improving Impact

Removing problem areas is only part of structural editing. The other goal is to use editing to focus the message of your work and improve its impact.

Balance

Let's consider how you direct the reader's attention. Everything in your final paper must be relevant, but not all the points you deliver will be equal in terms of their contribution to the conclusions. The challenge is to assign a proportion of the word count to the contribution each individual part makes to your wider message. This is a similar concept to deciding which figures are appropriate for the main body versus the supplemental, but now you are making the same decisions concerning the entire information flow.

In your first read-through, try to identify any points where the balance is not where you need it to be. These will either be long passage of text that aren't of major importance and should be shortened or critical pieces of information that don't have enough emphasis and need bulking out. In my experience, most new writers tend toward overwriting in first drafts so, more often, you will simply be looking to identify places where multiple sentences can be combined to deliver same material. In areas where you want to increase emphasis, without adding detail, you can do this through manipulating your paragraph and sentence structure. Adding a short sentence or paragraph between longer bodies of text can change the balance and rhythm to draw attention to a specific point. We'll come back to these topics in the next couple of sections.

Establishing relevance

In your structural edit, you will cut out irrelevant material, leaving just the relevant stuff. All good. Now you need to ask the next question "will my reader identify *why* this information is relevant?" I'm a firm believer that relevance should be explicit rather than implicit. Don't assume too much from your readers.

In short format writing, this should be less of a problem as readers will assume that everything is directly relevant, as it should be! You can rely on the reader remembering something you mentioned before when they read onward. If you are ever concerned something isn't obvious, all you will need is to add one clause or a short sentence to clarify the point.

In extended format writing, as in a thesis or dissertation, you will be dealing with many points and in greater depth than is possible when working within tight word limits. The potential problem areas are, unsurprisingly, introductions and the literature review. Your examiners will expect you to fully establish the basis for your work from the literature and to discuss thoroughly the implications of everything you have reported. This does not mean that you should write everything you know; everything you write should still connect to the overall message and points about repetition and balance hold true. What it does mean though, is that you will be considering more things as being relevant. This puts a greater onus on you, as the writer, to make sure that the reader knows what they need to take from each part of the text.

An easy way to help the readers is to include a short preamble to the different section that looks forward to where the story is going. "In this section, ...". Choose your phrasing to make everything as clear as possible. For longer sections, especially those that are dense with information, it can also be valuable to include one or two short phrases toward the end of the section to re-establish or emphasise specific aspects of direct relevance. Using both the top and the tail might be overkill. As always, try to put yourself in the mindset of a reader who is naïve to the specific subject but with the appropriate background level of knowledge and provide what they will need.

Examples: establishing relevance

Example 1: Forward connections to later in a subsection.
"… It is clear from these published works that although many proteins are associated with focal adhesion sites, a core subset of these proteins have been identified as marking the different maturation stages. We* have used these findings to focus our investigation onto these core proteins. Specifically, ….."

Example 2: Thesis subsection preamble.
"Later in this thesis, I* will describe the generation of a new in vitro model of arthritis. To put this new model into context, I will now detail the strengths and weaknesses of existing in vivo models …"

Note that these examples use first-person phrasing. Some supervisors may not like this style in formal writing. If necessary, use the third person, e.g. "This thesis will describe…" instead. Stay consistent with whichever version you choose.

In your first round of edits, focus on the "segues", the connections between the elements of your paper.

Flow

A primary consideration during this first round of editing is checking whether you have connected all the parts of the story. The more complex the material, the greater the requirement to use explicit connections. It should feel natural to progress from one paragraph to the next. If you find that the narrative jumps, then you have to ask yourself whether the problem is the link or the order of the material. Fixing the problem might simply take adding a forward link in your end of paragraph wrap sentences but, if you that doesn't work, you might find it is better to try switching paragraph order.

One area where new writers often lose the rhythm of their story is in the results sections of manuscripts, with the problems occurring in the transition from one subsection to the next. At the editing stage, look closely at the connections you have used (the segues). Have you established what drove you to ask the next question in your story? Have you been explicit in detailing what your objective was in the next set of experiments? Generally, if it is not immediately obvious, then it should be briefly explained.

If you are struggling with transitions, it could be because you are trying to stick to an ineffective storyline. Do not be afraid to move things! If figure three helps to explain figure two, then swap their order and have another go.

Example transition phrases

"Next, we asked......"

"To gain insight into..., we..."

"As it has previously been shown that ...... (+refs), we hypothesised that ...... was due to ...... To test this,..."

8.2 Editing Stage II - Paragraphs

Once you are happy with the overall content and primary structural elements, it is time to edit at the individual paragraph level.

Paragraph Length

Let me start by making it clear that there are no formal rules about paragraph length, it is always dependent on the content. Each paragraph should be complete but also discrete in terms of making a definitive single point. However, you can dramatically change the effectiveness of your writing, simply by reviewing then adjusting paragraph length to suit what you want to say.

As a **guide**, in formal science writing most paragraphs in the introduction or discussion of a manuscript and throughout a literature review should be around 150-180 words. If your paragraphs are much longer then you are likely to be rambling, adding non-relevant detail, or trying to deliver more than one key point. If your paragraphs have a tendency to be much shorter than 150 words then likely you will are not fully expanding on the point you are trying to make or have split things up too much and may need to combine the mini paragraphs that contribute to the same message.

Big Tip
Pay extra attention to paragraph with fewer than 100 words or with more than 200 to see if they can be improved.

If you are writing for the general public, the average paragraph length should usually be slightly shorter; approximately130-160 word on average. The overall structure of each paragraph should remain the same (topic, token, link, token, wrap), but the complexity of the individual sentences should be less, and they too will be shorter. You might have noticed that I have chosen to use relatively short paragraphs throughout this book; this was a deliberate decision to make the text less dense, easier and quicker to read.

Although it is appropriate to use longer or shorter individual paragraph where conditions demand it, one effective way to start your edit is to look at the length of each paragraph and flag those that are either below 100 words or over 200 words for further attention.

Short paragraphs (under 100 words).

For short paragraphs, ask yourself; are trying to make a point that is large enough to warrant a whole paragraph, or should it be combined with the paragraph before or the following one? If the paragraph is below 50 words, then you definitely have not said very much, and you usually do need to expand or combine. There is an exception, if you are making a deliberate choice for added impact (see rhythm section below). If it is slightly longer, 50-100, then consider whether you have got a clear topic and wrap sentence and are your token sentences fully developed? Try to be objective and critical so you can deal with any of the issues identified. If you are satisfied that these points are all satisfactory then it is probably OK to leave it as is.

Be aware that if you have lots of mini-paragraphs back to back, then it will feel like you never expand on any point, your work will feel superficial and the disrupted paragraphs will lack rhythm.

XXL paragraphs (over 300 words).

Extremely long paragraphs are harder to read, and, importantly anything in the middle of it is likely to be missed. For these reasons, it is almost always best to split any 300+ word paragraphs. If you realise that the long paragraph covers more than one central point, then breaking is easy. Identify which token sentences are associated with each point then split them up. Once you have decided on the cut point, add a topic and wrap sentence to each new paragraph, and you should be back on track.

If you cannot split the paragraph, then you have some serious editing to do. Identify what each sentence contributes to the message of the paragraph. You will need to remove some whole sentences, combine others and be ruthless in parsing down the length of the remaining ones. If you identify anything that isn't adding anything of value, then it must go. Delete is your friend. Be bold; your writing will be better for it.

XL paragraphs (250 to 300 words).

At the 250 to 300 words length, the easiest option is, again, to split into two separate paragraphs. If you have a general tendency toward writing long paragraphs, then it is probably a sign that you are also writing long sentences, or using two or three sentences to say something, which could be covered in one. Try to identify these traits and actively work to correct it. An important point to remember is that in science writing the token sentences frequently should be delivering the information from multiple sources, indeed it is not unusual for one phrase to cite three or four primary references. If you find you are writing a lot of single reference sentences, then look for whether this is the reason your work has become bloated.

L Paragraphs (200 to 250 words).

Every so often it is fine to have a slightly long paragraph, but if you have a string of these back to back it can make a passage hard work for your reader. It can feel heavy. Use the excess length as an indicator that you need to edit for brevity. You could cut one or more sentences or combine, but likely all you need to do here is cut excess words that aren't adding to the story. Some of the comments in the editing at the sentence level section below should help you with this.

Rhythm

In addition to the individual paragraph length, look at each paragraph relative to the structure nearby. Where you have multiple short paragraphs together, your writing will feel jumpy. Your work might read better if you combine some of those paragraphs to smooth out the journey. Similarly, a string of long paragraphs back to back will make a section of your work very heavy. You should consider breaking up one of the longer stretches of text to improve readability.

The point about is about reader experience, but changing paragraph length can actually be really effective way of selectively increasing impact in specific areas. A deliberately short paragraph can carry extra punch when surrounded by longer tracts. In contrast, but bizarrely similarly, deliberately increasing a specific paragraph can be used to emphasise that there is a weight of evidence to support the point. Strategic rhythm changing only works like this when it *feels* different to the reader, when the normal flow is disrupted. Don't worry about this too much at first, but as you develop your skills and confidence as a writer you can start playing with breaking paragraphs at different points to see what it does to how the message is received.

Structure

Topics and wraps

I've made this point a couple of times already, but it is crucial, so this is another reminder! It's time to re-read all your paragraphs topic sentences again and make sure they are working effectively. An excellent way to test whether you have effective topic sentences is to read *only* the first sentence of every paragraph to see if that would be sufficient to understand the whole story. This is a process known as *reverse outlining*.

Each paragraph should end with a "wrap" sentence that completes the point and provides a connection to the next paragraph

Next focus on the wrap. Read every paragraph and make sure there is a forward-looking wrap sentence, to round out the paragraph and lead to the next. In my experience, it is often the wraps that are missing from new writers' work, and leaves paragraphs ending very flat. It is worth spending a little time checking you haven't missed any.

Big Tip

Don't start a paragraph with a caveat, a definition, a difficulty, or methods issue.

Links and transitions (within paragraphs)

The body of each of your paragraphs will contain multiple discrete points (token sentences), with each relating to and building upon the topic sentence. The middle sentences of your paragraph tend to be dense with information and can end up as a boring list of factoids. To avoid this, you should think about what you are trying to deliver and consider using one of the transition or link phrases from the box on the net page. Each of these link phrases can help soften the hard facts and improve the rhythm of the paragraph, which, in turn, will make your work easier and more enjoyable to read.

Transition Phrases

- **Time/Order:** firstly, secondly, thirdly... finally. To begin with...
- **Continuing the same thought:** Moreover, furthermore, additionally, as well.
- **Contrast and Comparison:** in contrast, on the other hand, however, in comparison, comparatively.
- **Explanation:** consequently, in this way, because of this.
- **Support or contradiction:** in the same/opposite way, in agreement/disagreenebt with, this concurs with, on the contrary.
- **Cause/effect:** consequently, therefore, as a result, it follows that.
- **Generalisation:** largely, typically, generally, as a rule, frequently, in the clear majority of, mainly, mostly.
- **Conditional or concessions:** In this case, this is restricted to, notwithstanding, despite this, even though.
- **Emphasising or adding opinion:** surprisingly or unsurprisingly, naturally, unexpectedly, fortuitously, unfortunately, clearly, without a doubt.
- **Example:** for example, for instance.

Information Flow

Maintain a consistent viewpoint

During editing, check that your paragraphs maintain a consistent frame of reference. Pick a character or subject and a point of view depending on whose story you are telling, then stick with that viewpoint throughout the paragraph. When I say "character" here, it doesn't have to be a thing, it could equally well be an action or activity, but often your readers will enjoy your writing more if you frame the narrative around a noun. Maintaining a single viewpoint helps your writing stay focused and is easier for the reader to absorb. To maximise clarity, try to keep characters or subjects short and concrete.

Example: maintaining a viewpoint

Draft:
"During the first 24 hours of embryonic development, migration through the tissue takes place on the part of the progenitor cells. This migration occurs more quickly than the movement of other cell types travelling on the same path."

Edited:
"During the first 24 hours of embryonic development, progenitor cells migrate through the tissue. These cells migrate more rapidly than other cell types travelling on the same path."

> In the first version, the subject is *"migration through the tissue,"* which is fine in terms of consistency; however, changing the subject to *"the cells"* and presenting the information from their viewpoint improves the readability.

Progression: A to B, B to C, C to D.

Just as for the flow of the whole story, you should also consider how you build up ideas within each paragraph. When you are editing, check that you are building in the correct direction and that you lead from one idea to the next. Some paragraphs are more likely than others to have problems. The ones to watch for are where you change scale. Make sure you are consistently going in one direction throughout the paragraph; either big to small (in the introduction), or from small to big (in your discussion).

Example: Progression

Draft:
"Laminins and collagen IV assemble into basement membranes. Each laminin is a heterotrimeric protein."

Edited:
"Basement membranes contain networks of collagen IV and laminins. Each laminin is a heterotrimeric protein."

In both options, the sentences are individually fine. However, the different arrangement of the edited option helps lead the reader to progress through from A to B and then from B to C, in this case, going from big to small.

Avoid repeated changes in direction

Connections like "but", "however", "yet", "in contrast", or "on the other hand" act as a change in direction. Your reader can get lost if you include more than one of these type of phrase within your paragraph. If you find times where you have done this, try rearranging the sentences so that you group all the thoughts that are going in same direction and deliver them together. This will mean that you only need to

A paragraph with many changes in direction is likely to leave your reader confused about your message

change direction once. Conversely, if you have three or more alternatives, consider presenting them as a list with firstly, secondly, thirdly, etc.

Example: changing direction

Draft:

"These data could be interpreted as transcriptional activity having increased in cells treated with the drug. However, the same data could also imply that mRNA turnover rates have decreased, perhaps through reduced miRNA expression. Yet, the findings that miR-6, -16, -23 and -146 are all unchanged suggest that this is not the case. On the other hand, changes to miR 243 could account for at least some of the observed mRNA abundance changes."

Edited:

"These data could be interpreted as transcriptional activity having increased in cells treated with the drug. Consistent with this, many miRNAs are unchanged in the drug-exposed cells, which suggests that it is not a widespread change in mRNA turnover that is driving the transcript level changes. However, one miRNA, miR 243, displayed a large, potentially biologically significant change, and as such could account for at least some of the mRNA differences. "

Same information, same points made, but in the edited version, the sentences are grouped based on which interpretation they support.

Old to new, simple to complex

When you are introducing new concepts, pay attention to the order of your sentences and the clauses within those sentences. Deliver first the material the reader is likely to know already and only then introduce the "new" information. Structuring the information flow in this way means that the reader will not have to double back and re-read the start of the sentence or paragraph to understand the points you are making later. You always want your reader to get all the information from a single read-through.

This concept is relevant in all forms of writing, but it is particularly crucial when writing for lay or non-specialised audiences. In these cases, you will inevitably find that more of the terms you are introducing will be new and this means that need to pay close attention on how you bring in each new concept.

Example: Old to new

Draft:

"Haemostasis, inflammation, proliferation and remodelling are the four stages of wound healing. Haemostasis involves the formation of a clot within the wound environment, which stops further bleeding. This sparks the inflammatory response in which neutrophils and macrophages are sequentially recruited to the wound bed where they attack bacteria and clear debris. Activation of fibroblasts then occurs,

---

marking the transition to the proliferative phase. The fibroblasts divide and begin to secrete collagens and fibronectin which fill the wound bed as a provisional matrix. Contraction of this matrix by the activated fibroblasts also occurs, reducing the wound width. Also, during the proliferative phase, reepithelialisation occurs. Epidermal keratinocytes begin to divide and ultimately migrate over the provisional matrix until the wound is closed. Once the wound has closed the final phase, remodelling, can occur. Here, excess cells undergo programmed cell death, the provisional matrix is replaced, and realignment of collagen fibrils occurs in a process lasting many years."

Edited:

"Wound healing is a multi-stage process involving four distinct phases. The first phase, immediately following injury, involves stopping the bleeding by the formation of a blood clot, and is termed haemostasis. During the second phase, inflammatory cells are recruited to the wound and clear debris and bacteria. This inflammatory phase involves neutrophils first followed by macrophages. Once the wound is cleared, the third phase, the proliferative phase, can begin. This primarily involves activation and division of the fibroblast cells, which produce a provisional matrix of proteins, including fibronectin and collagens to fill the wound bed. The activated fibroblasts pull on this provisional matrix and contract the wound. At the same time, cells in the outer layer of the skin, the epidermis, begin to divide, then migrate over the provisional matrix in a process termed reepithelialisation. Finally, with the wound closed, remodelling can begin. During this maturation phase, excess cells die through programmed cell death, and the provisional matrix is replaced with collagen fibres, which are then slowly realigned. This final phase can last many years."

Note that the edited version is slightly longer but it is also more forgiving to someone who doesn't already know terms like fibroblast, fibronectin, neutrophils, etc. The challenge to the writer is that because you already know all the content very well, the order that these things are delivered does not matter to you. However, you are not writing for yourself, you are writing for your readers. Try to think back to before you knew the details and identify any parts that are likely to be new to someone coming fresh to your work.

8.3 Editing Stage III: Sentences

Tenses

Let's start with areas where there are clear rules.

Past tense

If you are ever in doubt, it is usually safest to use the past tense.

Past tense must always be used for materials and methods sections and any aims or hypothesis statements in a manuscript*. Past tense is also what you should use for most of your literature review and introduction, almost all the results and the parts of your discussion where you refer to previous work.

- Introduction: "Although it has been demonstrated that X influences Y, the mechanism is yet to be determined", "Here we investigated the contribution of X on Z and its effect upon Y. To do so we..."
- Methods: "Sample Y was incubated overnight at room temperature in compound Z."
- Results: "These experiments revealed that X increased two-fold when treated with drug Y", "next we investigated...."
- Discussion: "It has been hypothesised that X influences Y via Z; however, the results presented here have established that Z lies outside this pathway and has no effect on either Y or X."

(aims and hypotheses are future tense in grant applications or project proposals)

Present tense

The present tense is used a lot less frequently but it can be appropriate in situations where the statement either will remain true or is widely accepted as being true as the result of extensive research.

- Introduction: "Patients with disorder Z are born with features of Y." (established by extensive results)
- Results: "these data demonstrate that X influences Y in conditions where Z is present." Note that using present tense here makes the findings more definitive. This can help give your writing impact and make you sound confident in your interpretation. However, you must be cautious to avoid making conclusions beyond those supported by your data.
- Discussion: "These findings suggest an alternative interpretation where Y causes disorder Z."

Future tense

Use of the future tense is much rarer, but if you make any next steps or other suggestive comments in the discussion, it can be appropriate.

- Discussion: "To test this will require the development of X", "It will be interesting to discover if W actually drives X."

Viewpoint

There used to be a widespread belief that you always had to write in the third person (it rather than we) in formal academic writing, and you might find some supervisors still adopt and prefer this style. However, writing in the first person is more engaging for your reader. Therefore, there has been a shift toward using "we" or "our" whenever it is appropriate. But when is appropriate?

You must be careful to differentiate between what is truly "yours" versus something that cannot be owned or which is owned by the collective. Essentially, you can only use the first person when discussing a thought process such as "we hypothesised that..." or "we interpret these data...", or in the "why" and "how" sentences of your results, "next we wanted to...", "to do so, we..." Elsewhere, in most of your paper, you should use third person constructions like "these analyses revealed that", "results in mouse models support these findings".

> **Big Tip**
> Limit use of "we" or "our" to a maximum of once per paragraph and only for thoughts rather than actions.

> **Appropriate first-person phrases:**
> "We hypothesised that, ..." [intro]
> "These data led us to ask next..." [results]
> "We found these results particularly surprising in light of recent work describing..." [discussion]

Don't use the first person in these situations:

Anywhere at all in the methods section! (always 3rd person)

When describing the data from experiments. The data are being shared; you don't own them anymore! Instead of saying "our data indicated...." Say "these* data indicated..."

On top of these restrictions, avoid overuse first-person constructions. If every sentence says "we did this... we did that," it can become wearing and will feel unprofessional. If you go too far, a supervisor might tell you to remove every first-person reference. A way to avoid this is editing your sentences so that you have no more than one "we" or "our" per any individual paragraph.

*Data are plural.

Sentence length and rhythm

There are no hard and fast rules as to how long each sentence must be. Just like for paragraphs, sentence length affects the rhythm of your writing and is an area where you can improve the *feel* of a block of text through careful editing. The effect sentence length has on how your work is read can also allow you to add emphasis to specific points, increasing

> **Big Tip**
> Varying sentence length will improve readability.

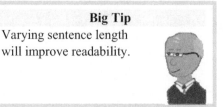

the impact of particular phrases.

General sentence-style comments

Too long = bad. I am uncomfortable giving specific guidelines on length as there are always exceptions, but let's do it anyway! Most sentences in a body of science writing tend to be about 10-25 words not including references. Sentences should be a little shorter when the audience is not specialised, 8-20 words. You will need some longer sentences and sometimes shorter sentences will help, be aware of your general style and tendencies. There are good reasons for not making a habit of using overly long sentences; research has shown that as sentence length increases, comprehension decreases. You want your wiring to be understood! Flag any sentences that are overly long or short for special attention.

If too long sentences are bad, then why do I not recommend using short sentences all the time? Firstly, quite simply, most things you are writing about are too complicated for single clause sentences. You will need a second clause to provide context or the caveat that is required to be precise and accurate. Secondly, your writing will feel very jumpy if every sentence is short. Each sentence break says to the reader that they should pause, that they should take a breath, even if they are not reading out loud. Variation is the key. Don't use lots of short sentences back to back. Instead, use short sentences as a mechanism to add emphasis.

Example: Changing sentence length for added impact
Draft:
Anabolic steroids are believed to produce muscle hypertrophy, but although it is generally assumed that the evidence underlying this belief is strong, this is not the case.

Edit:
Anabolic steroids are believed to produce muscle hypertrophy, and it is generally assumed that the scientific evidence underlying this belief is well-founded. However, this is not the case.

The sentence break adds emphasis to the edited version, gives it more punch.

Drawing breath

Part of your rhythm consideration is how you break your sentences into sub-clauses. If a period/full stop indicates a point where the reader stops, then commas, colons and semicolons are all mechanisms where you can introduce shorter pauses that allow the reader to "draw breath". But which one should you use and when?

Commas, colons and full-stops are indications to your readers of points to break and catch their breath. Try reading your work out loud. If you need to have an oxygen tank nearby, then it's likely you need more punctuation.

Commas, colons and semicolons

Comma

Use a comma to let the reader *take a breath*, to indicate a sub-clause, or to avoid ambiguity:

"He kissed the girl, who slapped him."

In the second case, without the comma you would not know if it was the girl who slapped him that he then kissed or if the girl slapped him after the kiss.

Semicolon

Use a semicolon to let the reader take a *longer* breath; use when there is something more to say about the same topic:

"The clinical signs typical of muscle damage are delayed stiffness and soreness; these are associated with elevation of serum enzymes.'

"Positive reinforcement can be a tool for eliciting behavioural change; it is widely used in human studies but less effective in cats.

Semicolons are also used when using a "however" in the middle of a sentence to start a new clause:

"Climate change is supported by experimental data; however, some people still debate the role of mankind in driving this change."

Colon

Use a colon where you want to indicate a breath *with an expectation*. Use one to indicate that you will expand a statement or introduce a quotation or list:

"Understanding the contribution of the circadian clock to biology has many potential applications: timing of surgery, calculating drug efficacy, identifying strategies to overcome jet-lag etc."

Sentence length: cutting out wordiness

Writing succinctly will not only help when you have to meet word limits, such as in your abstract, but will also make your general writing more enjoyable. In the editing phase of your paper preparation, aim to spot redundant words, or unnecessary repetition and cut them out.

Editing out unnecessary words
Delete sentences that convey no meaning:

"There is a lot of evidence from many workers on the actions of these various factors."

Pointless fluff! Delete, delete, delete.

Combine sentences that convey multiple aspects of the same point:

Draft:
"There was a significant difference between the experimental group and the controls (one-way ANOVA, Bonferroni post hoc test p<0.05). Specifically, the drug-treated cells were two times larger than untreated cells (mean difference 2.0 SD 0.4, p<0.05). Drug treated cells were also 2.2 times larger than the vehicle-treated group (mean difference 2.2 SD 0.2, p<0.05)."

Edit:
"Drug treated cells were two times larger than untreated cells and 2.2 times larger than vehicle-treated cells (mean differences: drug to control 2.0 SD 0.4, drug to vehicle 2.2 SD 0.2, p<0.05 for both groups, 1-way ANOVA Bonferroni posthoc test)."

Contract phrasing to focus on the delivery:

Draft:
"Sampling errors were reduced by the measurement of duplicate samples and the calculation of the mean in each case."

Edit:
"Means were calculated from duplicate samples."

Using sentence structure to control emphasis

We briefly mentioned above that a short sentence surrounded by long sentences can be used as mechanism to increase focus on whatever is in the short sentence. Here I want to discuss situations where you have two points that you need to deliver in a single sentence but where you want to put greater emphasis on one part of the sentence over the other. An example situation could be where your study has shown that a new drug treatment is safe for most people but there were some side effects. It would be wrong to fail to mention either of these points, but, depending on what message you want to convey, you might want to emphasise one part of the other. To do so, you can take advantage of one or more of the six emphasis rules.

Emphasis rules

In each example below, the same message is being delivered. Compare the two sentences and see if you agree that the "rule" holds true.

Greater emphasis is placed on the *primary clause.*
The drug is safe for most people, although there may be side effects.
Although the drug is safe for most people, there may be side effects.

Greater empahsis is placed on the clause at the *end of the sentence.*
Although there may be side effects, the drug is safe for most people.
The drug is safe for most people, although there may be side effects.

Clause length can intensity clausal order effects.
Although there may be side effects, the drug has been rigourously tested and is safe for most people.
Although rigourous testing have revealed that there may be side effects, the drug is safe for more people.

Least emphasis is placed on a subordinate clause *within* a primary clause.
There has been rigourous testing of the new drug and, although there may be side effects, it is safe for most people.
There has been rigourous testing of the new drug and, although safe for most people, there may be side effects.

Emphasis can be enhanced through repetition.
The drug has been rigourously tested in multiple studies and, while there may be side effects, it is safe with no adverse reactions for most people.
The drug has been rigourously tested in multiple studies and, while is is safe for most people, there may be side effects which can be severe and include the following features..

Emphasis can be amplified using modifier words.*
There has been *rigourous* testing of the new drug and, although there may be *rare* side effects, it is *completely* safe for most people.
There has been *rigourous* testing of the new drug and, although *generally* safe for most people, there may be *pronounced* side effects.
*be careful that changes to your sentence wording doesn't change the meaning!

Emphasise the real-world meaning rather than statistical significance

While we are talking about emphasis, time a reminder about what to emphasise! Remember that the reason you are writing the paper, and the reason you did any individual experiments is to deliver the answer to a real-world question. Although statistical statements are important in telling readers how confident you can be about your findings, the stats aren't the reason you did the experiment. On that basis, they should not be the focus of the sentence.

Edit your sentences so that the primary clause details the major finding and put the statistics in the secondary clause or, better still, in parenthesis. I mentioned this point in the results section but wanted to bring it up again here as it is relevant throughout your paper and fits with the idea that you

If you need to repeat something important, make sure you deliver the message using different phrasing

should now be editing to improve impact, not just correcting typographical mistakes.

Example: editing results to emphasise biological importance

Draft: "There was a statistically significant difference between population 1 and population 2.

Edit 1: "Population 1 was significantly larger than population 2."

Final: "Population 1 was almost two times larger than population 2 (mean difference 1.9 SD 0.2, $p<0.01$).

Active or passive voice

Scientific writing is often recognised as being difficult to read, dense and boring. Part of these criticisms come from breaking one of the core writing "rules": use of the passive voice. Or, perhaps that should be "passive voice is used in scientific writing". If using the passive voice isn't considered to be good writing, why is it so common?

Let's quickly deal with the fundamentals. Whenever your readers start reading a sentence, they want to get organised. They want to know whose story the sentence is telling, the *subject*. They also want to know the person/thing responsible for an action, the *agent*. When you use the active voice, the agent and the subject are the same, whereas in the passive, agency is deferred or implied. One way to think about this is *agent::action::outcome* means active voice, *outcome::action::agent* is passive voice. Possibly easier is to look at the verb form, if it is "was or were" then the sentence is probably passive voice.

The active voice can be clearer, it certainly uses fewer words, and, importantly, it makes your writing feel dynamic. If you are writing for a non-scientific audience, in a lay summary or press release, then deliberately editing your work into the active voice is a good idea. However, using the active voice was/is frowned upon in academic scripts. Good news is that times are changing. Nature Publishing Group, for example, "prefer authors to write in the active voice… as experience has shown that readers find concepts and results to be conveyed more clearly when written directly." It is amusing to note that "to be conveyed" is a passive construction in that statement, we all find it hard to break the habit!

There are four reasons for why, where and when you *should* use the passive voice:

Objectivity

In science writing, the things that we are writing about don't have agency. The story we are telling is the story of the cells, the molecules, the proteins, the experimental organisms. It is not about who did the work but rather it is about the results of that work. It isn't even usually the patients' story, but rather the story of data about the patients. Passive voice allows you to maintain a consistent perspective and, importantly, focus your readers' attention onto the correct place. Passive voice can imbue your writing with a sense of objectivity that otherwise might not be there.

Continuity and Flow

Context matters. Compare these two sentences:

Active voice: LaNts regulate the proteolytic processing of laminins
Passive voice: Laminin proteolytic processing is regulated by LaNts.

Both sentences are absolutely fine on their own, but which would be best depends on the context. The active version, would be used in a paragraph that is about LaNts. In contrast, the passive voice option would be in a paragraph that is focused on laminins.

As a general rule, you should change viewpoint infrequently. At the paragraph level, the delivery of the overall message is much clearer for readers if we maintain a consistent viewpoint. Even better, if you can maintain the same viewpoint throughout the document then your narrative will feel connected. Using passive voice might allow you to stay consistent.

Accuracy and Precision

Making the scientist the agent in the sentence isn't the only way to convert a sentence to the active voice, you could, theoretically, assign agency to the cells/proteins. However, the meaning can change. Compare the following.

Active voice: Elevated netrin-4 expression levels are associated with vascular disruption.
Passive voice: Elevated netrin-4 expression levels have been associated with vascular disruption.

"Are" has changed to "have been", and this has dramatically influenced the message. The active version is confident and definitive. It doesn't leave any room for doubt. In contrast, the passive version is more precise but also more restricted. It allows room for alternative interpretation and for other data to reveal something different. Choosing between these types of construction allow you, as the writer, to differentiate between the strength of inferences. This is a powerful tool that you will use again and again in your literature review, and discussion sections. Your language should agree with the quality of supporting evidence. I would expect the active sentence to be supported by three or more reference (and possibly even a meta-analysis) whereas the passive voice might have just a single citation to support it.

> **Active voice:** Laminin α3 bound to β4 integrin with high affinity.
> **Passive voice:** β4 integrin was bound by laminin α3 with high affinity.

The strengths and limitations of your experimental design should be reflected by the language used to describe it. The active voice here is stronger, but this is a situation where the origin of the data determines the answer. If this sentence was describing the results from an assay where one of the proteins was immobilised on a substrate and then the other protein added in solution over the top, then the sentence construction would have to reflect that design (passive voice, was bound). However, if the same experiment was repeated with the positions reversed then the active voice could become the better option.

Rules (materials and methods)

The fourth time to use passive voice instead of active is when there are official rules and you can't change them! In materials and methods sections, the only potential agency in methods is the "we" as the scientists doing the work, but the rules are that you must write methods sections in the third person. It is possible to give agency to things, but you can only assign agency where it is accurate. Experimental animals simply don't have agency if it is the scientist who is in control; "mice were mated". Even human subjects in an observational study are still being observed! Be aware, that the passive voice for methods rule continues to apply when you are using any rationale statements. You must say something like "Conditions were selected…" rather than "We selected…". Note that rationale statements aren't usually required in a methods section. If you need them, they often work better within the results section.

When can you use active voice?

If active sounds better, we should look for opportunities to use it where we can. The first thing to do is consider any sentence where you have used weak verbs like "is" or "have" and consider converting them to active forms. Unfortunately, one of the downfalls in writing for scientific audience all the time is that the tendency to write in the passive voice becomes ingrained. It's become almost my default position, and I have used lots of passive voice constructions as my examples.

The big question is about ownership and agency. In what situations is ownership *relevant* to the statement. Usually, it is any time when you are describing a thought process.

> **Active voice:** We hypothesized that
> **Passive voice:** It was hypothesized that

In this situation, changing to passive actually makes it sound like someone prior to the study had come up with hypothesis. The second sentence probably should carry a citation, whereas the first one is clearly the experiments.

> **Active voice:** We used confocal microscopy to determine the relative location of laminins and integrins.

> **Passive voice:** Confocal microscopy was used to determine the relative location of laminins and integrins.

In these situations, either is fine but here I would lean toward the active. Although it doesn't matter who did the work it doesn't distract from the narrative and there can be benefit from being a little more dynamic and accessible.

> **Active voice:** "They investigated cell population X."
> **Passive voice**: "They performed an investigation of cell population X"

The last example here is the sort of sentence you might write in a literature review or possibly when comparing and contrasting in a discussion section. Again, both options are fine, but the feel of the sentence is much better in the active voice. It is shorter, more dynamic, with the action in the verb.

Remember, editing is about doing the best that you can, not every "rule" can be followed every time. If you use grammar checkers on your word processing program or apps like Grammarly, sentences will be flagged that are in the passive voice (even this sentence is in the passive voice). These tools are a way to spot those sentences that it might be beneficial to change but definitely don't think that you have to change everything!

> **Don't use "our results", you can't own results!**

Author et al., sentences...

Do you really need to mention the author's name?

This point isn't about whether you should reference other people's work (you should, end of discussion). This point is about whether you should use the author or group's name as part of the sentence where you describe the work, e.g., "As shown by Smith et al...." It is a little about emphasis and a little about passive voice!

The potential problem with mentioning the author's name is the sentence is that you are highlighting *who* did the work. Doing so reduces the emphasis upon the true reason you have written the sentence. It is not wrong to highlight the author's names, but the same sentence could be more focused and, therefore, more impactful by only mentioning the author in the citation.

> ## Example: replacing "Author et al."
> **Draft: Author et al., adding no value:**
> "Using a mouse model, Yurchenco et al. 2018 investigated the role of laminin network assembly in disease. They discovered that muscular dystrophy occurs in animals where networks were incomplete."
> **Edit:**
> "Muscular dystrophy was shown to occur in a mouse model of defective laminin network assembly (Yurchenco et al., 2018)."

In addition to changing the emphasis, phrases centred upon *who* did the work don't say anything about *your* opinion. Indeed, you are doing the opposite; you are distancing yourself from the findings. Occasionally, that is will be what you want to do.

Big Tip

When referring to published work, it is the findings that matter rather than the name of the authors.

However, If I am marking your essay and I see that everything was delivered in a "Smith did this, and Jones did that" way, then I automatically think that the student has not thought about what is going on, they are just repeating what they have read. If the marking rubric includes evaluation of synthesis, then the paper filled with "et al.", won't score as highly.

When "author et al." phrases can help your writing; to present controversial or opposing positions.

For every rule there are examples of how breaking the rule can be beneficial. The situation for author et al. sentences is when you actually do need to connect the authors' names to their hypotheses or findings. This could be a paragraph in your discussion where you present two or more different opinions, then compare your study to the previous work and conclude about which model is supported by your findings. A similar approach could be to highlight similarities in different approaches, e.g. if a mouse model gave the same results as a zebrafish model, you could point out who used what. Here the author name may not matter, but the point you might want to make is that multiple *different* groups came to the same conclusion.

Example: when using the author's name adds value

Use #1; assigning a viewpoint or hypothesis to the source:

"Yurchenco et al., propose a model where network assembly requires interaction between three arms… (+ something saying if your data support/refute that model)

Use #2; presenting opposing opinions:

"Work from the Yurchenco and Hohenester groups strongly support the three-arm hypothesis, whereas a study from Odenthal et al. suggests an alternative model..."

Use #3; comparing your work to the literature:

"Recent work from Ahmad et al. indicated … however, our data point to an opposing interpretation."

Don't use "As shown in Figure X."

This point is similar to the et al., sentences. Again, you need to consider what effect the sentence construction has on the emphasis. If you say, "as shown in figure 1, X is 2x bigger than Y", the focus of the

Big Tip

Figures should support the text not the other way around.

sentence becomes on the *location* of data rather than on the *result* of the experiment. Remember that the figures are the evidence you provide to support a conclusion, your message should be clear without your readers needing to refer to the figures. For maximum impact, edit your sentences so that the first clause is the data and put the figure or table location in parenthesis at the end (Figure 1).

Example: referring to figures and tables

Draft:
"Table 1 shows that population 1 are more likely to suffer from the disease than population 2."

Edit:
"Population 1 are more likely to suffer from the disease than population 2 (table 1)."

8.4   Editing Stage IV: Word Choice

The next step I want to talk about is improving word choice. Don't worry; you are almost finished!

The big stuff

Accessibility

So far, I have written primarily about content structure and organisation, but there is no point having great content if no one can understand your writing. The biggest problem I encounter with new writers, somewhat surprisingly, is trying too hard to sound smart! The truth is, you don't have to use complex language. Indeed, your aim should be for your writing to be as accessible as possible. Whenever you have a choice between the phrasing used in regular lab conversations or using more complex alternatives with the same meaning, then select the more straightforward version. Simple and clear messages are always best.

> **Big Tip**
>
> Clear and accessible language will be more effective at delivering your message than fancy prose.

I have asked you to be precise in the detail of your methods and the reporting of your data. Now I am asking for accuracy and precision in language too. Using your word processing programme's synonyms suggestions may seem like a good idea at first, but you may subtly change the meaning by selecting a different word. You must use phrasing that accurately describes what you want to discuss. Sometimes this does mean that you have to compromise a little on accessibility in certain places.

Continual use of the same phrasing

You will find, at some point, that you have data from multiple similar types of experiment to present. The temptation is to use identical wording and the same paragraph structure to detail each set of similar data. If it works for one, it will work for all. Well, sadly, that's not quite true. If you copy and paste and change only the numbers it will be *boring* to read. I know it's a challenge, but make your writing more enjoyable by changing the phrasing wherever possible*. The good news is that when talking about similar data sets within the same body of work, you can reduce the amount of preamble and connections in the second and any subsequent time you are delivering related content, i.e. you will be able to deliver the same content in fewer words, not just different words!

There is less of a problem with repeat phrasing in figure legends.

Avoid cop-outs

Phrases like "It is well known that...", "It is well documented that...", "It is widely believed that...", "As everybody knows..." all mean the same thing: "I can't be bothered to look up the literature", if you have used any of those then it is best to delete them! Another commonly used fluff phrase is "a number", which means anything more than two. It is extremely vague and effectively worthless. Either be precise or use a different construction.

Common Mistakes to Lookout For

Here are a few areas where many writers struggle. My thinking here is that by flagging these points, you will know what to look out for whenever they come up in your writing.

Hyphens and Dashes

Hyphens are used to link two connected words that either don't make sense on their own or which would have different meanings if left unconnected. Use hyphens to avoid ambiguity; "A clip on lid" vs "A clip-on lid", "cutting edge topics" vs "cutting-edge topics". Also, use a hyphen whenever you are using units as adjectives "a two-minute incubation", "a three-month old mouse", but do use not for when the time units are nouns, "an incubation of two minutes".

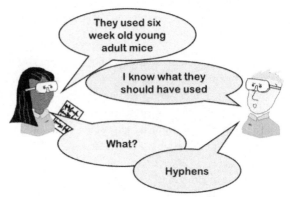

Hyphens can help remove ambiguity by identifying which parts of the phrase are connected:
Six-week-old, young-adult mice.

You might see dashes used on website to separate sub-clauses; however, in formal scientific writing that's the job of commas. Do not use dashes.

Parenthesis

In a scientific manuscript you will use brackets for references, for reporting descriptive or inferential statistics and for indicating figure locations. That is already quite a lot of sets of brackets. Therefore, my general advice is to not use any extra parentheses to indicate asides. As for dashes, it's better to put the parenthetical phrase within commas instead.

Indefinite articles: it, these, they, this

Make sure every time you use an indefinite article that there is no ambiguity about what "it" is replacing. If there is room for any doubt at all, then you should redefine. Two simple rules to employ:

- Always redefine your indefinite articles in a new paragraph.
- Don't carry an indefinite article across multiple sentences.

Example: indefinite article causing ambiguity:

"Raised interocular pressure is a major risk factor for glaucoma. **It** is caused by either increased production of aqueous humour or blockage of drainage through the trabecular meshwork."

Does "it" refer to interocular pressure or glaucoma?

Check your writing, if there *could* be doubt then change the sentence

---

Don't say prove, show, or fact

Part of our scientific training is to be cautious with the interpretation of data. We use statistical tests to indicate how confident we can be for an individual finding and those tests tell us how we should write about our findings. A low p value means we are confident in the result, but we are never 100% sure. A single experiment can never prove something. Quite simply, this means we can't usually use absolute terms like "prove", "show" or "fact" both when describing prior literature or new results. There really needs to be an enormous wealth of evidence before you can even talk about Theories! It is best to soften your language to "indicate", "suggest", "demonstrate" or "reveal".

Data can either *support* or *refute* that hypothesis, they cannot prove it to be true. Published works are not necessarily "facts", they are "findings."

Example: prove, show, fact

Draft:

"The findings prove that X causes Y."

"These data show that our hypothesis is correct."

Edit:

"The findings suggest that X causes Y."

"These data support our hypothesis."

Significant meaning important

In general conversation, "significant" is used for emphasis, as an adjective to mean notable, major, large or noteworthy. Politicians and the journalists also tend to overuse significant as a method of avoiding being specific. Using a subjective term rather than any objective measure gives them flexibility to deny anything!

In scientific paper writing, whenever you write significant the reader will think that you are talking about

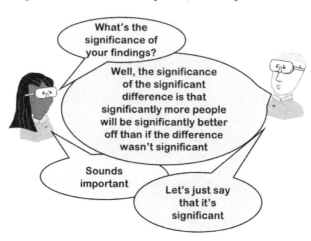

Significant has lots of meanings, make sure your phrasing doesn't introduce any ambiguity

inferential statistics, unless you make it clear that you mean something else.

Three easy strategies to avoid any ambiguities:

- Anywhere where you have written significant, change it to a different word. A better, more precise word! This is my preferred strategy.
- When you say significant in situations where you are citing statistical

information, remove the word significant and let the P value report the results of the statistical test. Refocus the sentence onto the real-world finding.

- If you want to leave significant in the sentence, clearly state whether you mean either "biologically significant" or "statistically significant".

While you are looking at your significance sentences, make sure you are using the words correctly. Significance tests can only give a binary result, either significant or not significant at whatever level you defined as the cut off (the α level). A value can

Big Tip

Absence of proof is not the same as proof of absence.

"fail to reach the threshold for significance", but it can **never** "trend toward significance" or any similar types of construction. State the P value as reported and use that P value to shape the way you describe your confidence in your results.

A reminder, if you test for a difference between populations and are returned a value that says "not significant" you cannot conclude that there is no difference. Your P value gives you a probability of sampling error being the cause of observed differences. If you want to test for a absence of difference, you will require a different test.

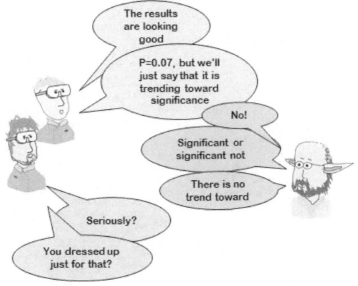

Comparisons

Two points to watch out for:

- Sentences with relatively or comparatively *must* also contain what is being compared against.

  "The gibbon has relatively long arms" (relative to its legs? its body? to other primates? to aphids?).

  "The chances of a mutation proving lethal at this stage of development are comparatively great" (compared to other stages? to the chance of being struck by lightning?).

302

- "X is larger than Y" NOT "X is larger as compared to/with Y", or "larger when compared to/with", "larger in comparison with"
  It is technically OK to say: "A comparison between A and B reveals that A is the larger", but it is a longer phrase with no extra value.

Only and Just

In speech, you can put "only" or "just" almost anywhere and change the meaning by placing different stress upon the word in how you say the sentence. In writing, you can't do that.

Example: "only."

Compare:
'This substance is only secreted by mature cells.'
With:
'This substance is secreted only by mature cells.'

Sentence 1 means that the substance is not *made* by the mature cells; the mature cells only secrete it. Sentence 2 means that other cells (immature cells) do not secrete the substance. This sentence does not say anything about which cells are making the substance. Which one is correct? It depends on the substance, if only we knew what it was!

Both

Usually, you will have included "both" in your draft as a method to emphasise that what you are saying refers to more than one population, but you must be careful.

If the first noun after the both is a plural, it might be unclear whether you mean that you had two of the first noun of the pair or whether you mean the entire pair. The chance for confusion rises when you are talking about things that come in pairs, for example "both parents and children", could be talking about a specific set of parents or a more generalised description of all parents. If there is any chance of ambiguity, you should edit the sentence. Sometimes adding "the" to the noun helps. Other times, most often, deleting the both is actually the simplest and best solution.

Example: "both"

Compare
Basement membranes contain both laminins and collagens.
With
Basement membranes contain laminins and collagens.
Or
Basement membranes contain members of both the laminin and collagen families of proteins.

In the original sentence it is not clear if you mean two laminins or many laminins. The edited versions remove that *potential* source of

---

> ambiguity. My preferred option is dropping "both" to keep the phrasing short.

If you add "both" to your sentence, make sure it
doesn't add ambiguity!

Commonly Confused Words

Data

Data are plural. The singular is one datum point. Make sure you use the plural constructions; "these data are".

Effect or affect

It is easy to get these two mixed up, so it is always worth double-checking.

Affect is usually a verb "The medicine affects blood pressure".

Effect is usually a noun "The drug effect was measured…". Occasionally effect can act as a verb, generally meaning to produce, e.g. "effect a change."

Use or utilise

This problem comes from trying to sound overly fancy. Utilise and use are not synonyms. Utilise means to use something for a different task than its standard use. For example, you would *use* a ruler to measure the length of something or to draw a line but might utilise a ruler to prop something open. Similarly, a PCR machine is *used* to amplify DNA, and a heat block is used to heat. Check every time you have used utilised and utilise use unless you should use utilise instead!

"A microscope was used to visualise transfected cells."

"The end of a 15 ml test tube was utilised as a holder for dissected eyes."

Less or fewer

Use fewer whenever you are talking about people or things in the plural (participants, cells, interactions etc.). Use less when you are referring to something that can't be counted or doesn't have a plural (time, brightness, proliferation).

"Fewer cells divided after receiving treatment one than following treatment two."

"There was less proliferation in the cell populations treated with drug one compared with drug two."

Additional times when less is appropriate are when numbers are used on their own or with expressions of measurements or time.

"Proliferation dropped from 90 % of cells per 24 h to less than 15 %."

"Expression lasted less than 20 h."

Various or varying

Varying means different each time.

Various means there was a range of different things.

Most of the time you don't want to use either option! It is better to be precise whenever you can. Words like "pre-determined", "pre-specified" or "defined" all sound like you did things with a plan and had more control over the experimental setup.

Continual or continuous(ly)

Use continual where something persists (recurrent, persistent, regular).

Use continuous when you mean "all the time" (constant, non-stop)

"There was a continual problem of staphylococcal infection."

"Patients were under continuous supervision during this task."

Occasional

Use occasional *only* for time-related phrases and not for spatial or frequency information. Replace "occasional cells stained..." with "A small proportion of cells stained..."

That, which or who

If the part of the sentence after "that" or "which" is a *necessary* part of the sentence, and you would not pause before it in speech, use 'that', with no preceding comma (i.e. for an essential or restrictive clause).

"We analysed the data that the experiment delivered."

If the second clause gives only additional information, and you would pause before it in speech, use 'which' (or 'who') with a preceding comma (i.e. for non-essential or non-restrictive clause).

'We analysed the experimental data, which was normally distributed."

That vs Which

Compare:

"Cells that had prominent lamellipodia were analysed."

"Cells, which had prominent lamellipodia, were analysed."

The first sentence means that only the cells with large lamellipodia were analysed, whereas, in the second sentence all the cells were analysed; the size of the lamellipodia is included as a description for the reader.

In both cases, although you know what you mean, a reader might not appreciate the subtlety in difference of meaning. To avoid any risk of ambiguity, it might be worth restructuring:

"Analysis of cells with prominent lamellipodia revealed..."

> "Cells displayed prominent lamellipodia. Analysis of all cells revealed..."
>
> Being clear and precise is always a good thing.

Infer or imply

Use infer when you are drawing a consequence from your findings. Use imply/implies when you are actively laying out the results.

"This observation implies that an inhibitory factor has been removed."

"We may infer from this observation that an inhibitory factor has been removed."

Substantive or substantial

Substantive refers to quality, meaning or solidity, whereas substantial refers to numerical abundance.

"Re-wording the study questionnaire substantively improved the data quality."

"The extra questions added to the questionnaire substantially increased the amount of data obtained."

I quite like using both together for added emphasis;

"We submit here a revised version of our manuscript, which we have expanded substantially and improved substantively in response to reviewers' comments."

8.5 Editing Stage V: Final Checks

This final editing process takes an eye for detail and needs to be done in small batches to stay alert.

Read it out loud

Your writing should sound pleasant to the ear as well as on the page. The added value in reading your work aloud is that it will help to highlight any rhythm issues. Overly long sentences or regions where commas or other punctuation are missing will all become more noticeable as you will run out of breath. Similarly, speaking through a section will help to highlight any stretches of short, staccato sentences where the flow is being interrupted.

Read it backwards

By this point you will have read all your sentences multiple times, you know what you *think* you have written. When you read the text now, you will complete sentences in your mind making it harder to spot any problems on the page. Try starting at the end of the document and reading everything backwards. At very least, mix up the order that you read the subsections. If you always read the conclusion or figure legends last, when your eyes are tired, you will miss more problems there than elsewhere.

Standardisation and abbreviations

Do one final run through to check everything is consistent. Units, abbreviations, Fig/Fig./Figure etc. Check all references are appropriately formatted in the text and in the reference list. Make sure your word usage is consistent in terms of hyphenation, numbers, capitalisation, and fonts. Do a final check of the figures (printed out) for fonts and lines consistency, and to be sure that nothing has moved.

Run the spell and grammar checker

Spell and grammar checkers are not perfect. However, spell checkers will identify some things that you really should not have missed. Many of the scientific words won't be in the dictionary at first, but it's still worth using the tools available to you to catch anything you can. Adding field-specific words to the library will mean the spell checkers get better each time you use them.

In addition to the built-in spelling and grammar checkers, there are several apps that could help your writing. Again, these *can* help, but you will not be able to rely on these exclusively, and sometimes you need to ignore their advice.

It can be quite embarrassing to discover that you cat has contributed to your manuscript.

It's never going to be finished!

You could endlessly edit your document, but there comes a point where you are making only superficial cosmetic improvements and your time would be better spent elsewhere. What is most important is that the message is clear and that your co-authors are all happy. Once you reach that point, pull the trigger and submit. Remember that if you are sending your work to a journal article, the reviewers, editor and copyeditor will all comment on your work, and you are likely to be editing again in a few weeks. Once it is finally accepted, you will also check the final page proofs before the article is published.

Chapter 9: Talks and Posters

9.1 Talks
Advice for giving an excellent science talk.
9.2 Posters
Present your work at scientific conferences with style.

9.1 Talks

Research isn't just about finding out new things; it is also about communicating those findings to others to allow the scientific community to advance. So far, we have focused on the writing aspects of science communication. However, there are many other mechanisms in which you can share your story, the most common being talks and posters.

There are many occasions when you are required to talk about your work. You will most likely start with informal talks to your peers at lab meetings or internal seminars (a great chance to hone your skills). As you and your projects progress, these talks will grow in importance. You will submit abstracts to conferences and hope for an oral presentation slot, and later in your career you will begin to be invited as a seminar speaker and to give lectures. You might also be asked to talk to the general public or non-academic specialists as part of an outreach event or pitching session. Being an excellent speaker will not make or break your career, but it can certainly help. It becomes valuable as you progress through your career as talks are a staple part of job and fellowship interviews.

Public speaking might fill you with dread at this time but don't worry; it gets easier with practice. Also, good news, the general standard of science talks is not all that high so you have a good chance of being able to shine! Any nerves or fear can be lessened by having a presentation that you are confident "works", knowing that you have prepared well and have practised. In other words, effective preparation can make a big difference.

Being able to deliver a talk in an engaging and effective way is a skill that needs practice.

Respect your audience

It isn't you who decides if your talk is good or bad, that decision is up to your audience. Just like writing is for the reader, everything about your planning and delivery for a talk should be with your audience in mind. It is they who have given up their time to listen to you; you need to reward them for that time; otherwise, they will become distracted, bored and frustrated.

Thank you

At the beginning and end of your talk, thank the audience for coming and for listening. Acknowledge that you respect them and value their input into your work.

Have a clear objective for the talk

Important Point

Your goal for the talk is to be *effective*. Being effective means that your audience are rewarded for listening to you by gaining something from the experience. In many talks your primary purpose is the delivery of information rather than being for entertainment. It is a good thing to be engaging, but never lose sight of the reason that you are presenting.

Your talk is for the benefit of the audience. Respect their needs and thank them for giving up their time to listen to you.

Step one in preparing a good talk is identifying the primary goal of the presentation. The most common reason for a presentation leaving listeners dissatisfied is when it lacks focus. Lack of focus could come from delivering too much or too little context for your audience or going off on an irrelevant tangent. Therefore, before you begin preparing your slides answer this question; if the audience were to remember only one thing about your talk, what would it be? This goal statement *should not* be a vague overview of what you want to talk about. Rather it should be as succinct and specific as possible. Although you personally might gain something from a presentation, your goal should always be related to how you want *the audience* to benefit. The goal, therefore, clearly will depend on the audience and the reason for the talk.

For a science talk to other scientists your goal will usually be to convey some specific new piece of information, much like when writing a paper. In these cases, it is good to start with that conclusion. Decide where you are going before doing anything else.

In more general terms, the goal of a talk might be to engage, excite, or provoke the listeners, to make them glad they came, to make them eager to read more about your work, or it could be that you want them to see how you would be a useful addition to their team.

All the suggestions hereafter help to make incremental improvements to individual parts of your talk but having beautiful slides and delivering them perfectly cannot rescue a presentation that has not been explicitly designed to suit its audience.

Checkpoints

Once you have a defined objective, you will be able to identify the pieces of information or steps required to reach that goal and, equally importantly, the things that are not needed.

Next, identify the key checkpoints that you need your listeners to absorb to reach the overall objective. The checkpoints you identify provide focal points for the different stages of your speech; the rest of your material should be designed to maximise the impact of these key points. The material you deliver should lead you and your listeners from one part of the speech to the next.

Identifying checkpoints allows you to break up the talk into small parts that are easier to focus upon. The number of discrete checkpoints required should reflect the talk length; around four or five points for an eight-minute talk, seven for twelve-minutes, and twelve to fifteen for a 45-minute seminar. You are likely to benefit from writing down your checkpoints. Indeed, early in the talk preparation process, you might find it helpful to prepare a slide for each of your checkpoints to act as a framework to build the rest of the material around.

The final checkpoint should always be a conclusion. The conclusion is what people will keep in mind and take away with them, and it should get extra attention.

Example: identifying checkpoints

Overall Goal of the talk

Convince listeners that the protein LaNt is a prognostic biomarker for breast cancer.

Checkpoints

What the listeners need to either know or need to be convinced by to reach the overall goal:

1. Why the need for better breast cancer biomarkers.
2. Rationale for choosing to investigate LaNt.
3. Establish that experimental approach is appropriate.
4. Establish that there is a correlation between high LaNt expression and incidence of metastasis is strong in discovery cohort.

---

5. Establish that there is a correlation between patient outcomes and LaNt expression levels.
6. Conclusion: LaNt is prognostic biomarker.

Talk Structure

Now that you have established *what* you want to deliver, you next need to decide how best to reach that objective while keeping the audience interested throughout. Let's look at the five sections most talks will need:

Structure
1. The opening (hook).
2. Transition.
3. Introduction, Rationale, Aims and Hypothesis.
4. Body (Results for a science talk).
5. Conclusions.

The opening

First impressions count. From the point when you first step up to the podium, you have about one minute where *all* the audience will be listening, fully engaged, prepared and willing to give you a chance to succeed. You need to grasp this opportunity and keep a hold of the audience. Engaging the audience early will give you the best chance of maintaining their attention through to the later parts. Lose them now and it will be an uphill battle to get their attention back.

Overall, your opening should answer their question "why should I tune in to this talk?". In practical terms, this means you should introduce the talk. That sounds simple, but it also sounds boring. The key to making it interesting is actually not putting the emphasis on the *content* but rather emphasise the *value* the audience will gain from listening to you.

The very first thing you say should be something to grab the attention of your audience, "hook" them now and reel them in later. So, how do you bait your hook? You have four main groups of options. Which one to choose depends a on the audience, and where you are going to take them.

The Hook – Grabbing your audience's attention
Once upon a time.
Tell a story that says why the topic affects life on a personal level. "I am going to telly you a little story about Annie who has a rare skin blistering disorder…"
Using your own experience here can work but remember that talks aren't usually about you, they are about the audience, so using a personal story can only wor if the listeners will be able to see themselves in your story. If you go down the story route, the ideal scenario would be to come full circle through the course of the talk,

returning to your opening anecdote again at the end to give a feeling of completeness.

A factoid.

Opening a talk with a statistic that shocks the audience into thinking or re-thinking about the topic area can be an effective option and is an approach that you will see used often. "Did you know that although rare diseases each affect fewer than 5 in 10,000 people, approximately 7% of the population will be affected by a rare disease at some point in their life."

Opening with a factoid can be a little risky if most of your audience already knows the fact that you intended to shock them with! If ot is widely known, then the effect will not be an attention grabber but rather an indication that the listener can switch off. For example, telling an institute that does research into ageing that the population is getting older is unlikely to grab anyone's attention.

A question that the audience cares about.

Identifying questions your listeners would like to see answered is usually quite easy when presenting primary data. The whole basis of the research you are presenting was likely to be a question that was worth answering. If you open with a question, then in your conclusion, you should aim to explicitly revisit the question and give the answer and a feeling of completeness..

Icebreaker

You can also start with some an amusing or interesting observation, which is not necessarily related to their talk content. This sort of opening hook can soften up the room and make the audience feel that they are going to enjoy the talk. If there was a lot of chat in the room beforehand or a lot of people entering or leaving just before you talk, then opening with something that isn't important to the grand scheme of the talk, can also be a useful strategy to change the feeling in a room and tell the audience it is time to listen. I almost always do something like this whenever I am giving a lecture in a teaching environment. My recommendation is that you should not use an icebreaker to *replace* a project-relevant hook. Use an icebreaker instead to grab the attention so that your *real* opening can have the greatest possible impact.

Data teaser

One approach that I really like is showing an exciting piece of data that will be thought provoking as my first slide. From my lab's work this is almost always something with visual impact that is easy to understand without context. I use an image to show that something exciting is coming and I signpost it as being a point where I solicit the audience to engage. For example, in a recent seminar I started with a panel of histology pictures from a new trangenic animal model and invited the audience to come up with their own interpretation of the phenotype before I told them anything about the model, the transgene or our hypothesis. Scientists like to think, so I gave them something to think about.

Did you notice one opening that you see all the time is missing from the list. "Hello, my name is…" Introducing yourself is not an effective "hook". The reason for the audience to care about your work is not because it is you that is delivering the material; it is because the content has value. Credibility is earned, it comes from what you say and the data you show not from your description of yourself. I prefer to give my name and contact details at the end of the talk, once I have convinced my audience that they should contact me.

The one time when it is likely to be beneficial to open with an introduction about yourself is during a job talk, but even then I would use a hook first and introduce second.

The transition: "But before I can tell you about that…"

Now you have "hooked" the crowd, it's time to transition into the meat of the talk. You might have heard the adage "tell the people what you are going to tell them, tell them, then tell them what you've told them." Fundamentally, this is reasonable advice, but there is a real danger of being boring and losing your crowd if the first step goes on for too long.

The transition should tell the people what you are going to tell them, *and why they need to know.*

In conference talks and seminars, after the transition, you will deliver the relevant background material needed to appreciate the value of your work to the field and the information needed to understand the rationale behind your study. A good option is to use a one or two sentence synopsis of where most of your story will be centred and then step back to telling the audience that you will explain what led you to carry out that particular study.

You might have been told to include a slide with outline of the talk. If you are thinking about having a slide that says "introduction, methods, results, conclusion", please don't bother! This sort of stuff is so obvious that the listeners will switch off.

> **Big Tip**
>
> It is better to have absolutely nothing on the screen than having something
> that says;
> "don't listen, this is boring."

The point of the outline is for the audience to see the journey that they will be going on. If the talk is relatively linear, you almost certainly don't need a slide; you could deliver the same content just by talking. However, outline slides *can* work as transitions if the story has some twists. Focus the words you say upon how the listeners will benefit from each section. Even so, don't labour this slide. If you use one, it should be short and focused.

Introduction / Background / Rationale

In conference talks and seminars, the research you will present will need some context to help your audience identify why it is important to them, and what was the thinking that shaped the research direction.

By far the most common problem in science talks is that the introduction contains too much irrelevant material or just goes on for too long. There is a big difference between a lecture and a research project talk. If you are giving a data talk, the *value* to the audience, the reason they have attended, is in discussing the new findings and how those data have changed the view of the world. Your introduction shouldn't be designed to teach anything, it should be tightly focused on establishing the context within which your work added value. Just like the intro to a paper, it is about "mapping the gap".

The challenge lies in giving enough information to bring the audience members who are new to this area up to speed while simultaneously delivering information rapidly enough that the experienced audience members don't switch off and become bored. Achieving this balance can be difficult, which is why it is essential to be focused. So, let's focus; what should an introduction contain?

Introduction

Just like in an introduction to a paper, you have a few things to cover:

Motivation

Why should the audience care about this work? Often this is covered by your opening remarks, but you might want to reinforce the key drivers behind the work again (briefly) to add emphasis. Frame the motivation in terms the audience will appreciate. For example, rather than saying "we wanted to know" change the emphasis to "it would be useful for the world/field to know".

Relevant background

You will need to deliver the *directly relevant* material that shaped the way you interpreted your data, or which defined how you asked your questions. To keep the audience interested during this part, make sure to also to tell them why *they need* that specific information. Often the people in the audience will know the background, so your job is to be selective about the information you deliver and present those pieces of information in a way that makes their relevance to your study clear.

Rationale

Thirdly, establish the basis for your hypothesis in clear terms. Alternatively, if your study has more of a translational lean to it, present the objective that you are working toward. This part is related to the motivation, but now you should frame it into smaller, shorter-term targets that you have addressed.

There are overlaps between these parts, so feel free to combine them into a single element, especially if you can reduce the length of the introduction by doing so.

There are no hard rules for how long your introduction should be; it depends on your audience. However, from my experience, new students tend to talk for too long is this section. In a conference-style 12-minute presentation, I recommend that you should aim to move into describing your study after two to three minutes of background. In a 45-minute seminar, you can justify a more extended introduction but make sure it is filled with things that are of interest to your audience, don't just tell them what they already know!

A final point on your introduction. At scientific conferences, you will probably be presenting your work within a session where the talks surrounding yours are on similar topics. This creates a danger that you will repeat things that other presenters have already said. Before you go to the meeting, check the program and read the abstracts of the other talks in your session to identify what you think will be covered. Then, on the day of your talk, be prepared to go faster or slower depending on what has gone before.

Think about your audience.
The length and content of your introduction
should be tailored to what they require.

The body

Most of your talk will be about delivering your data, the new findings. Your audience will appreciate and enjoy this section of your presentation more if you guide them through the journey, making the understanding easy for them. For each set of experimental results, you should tell the listeners the experimental question or hypothesis, how you addressed the question, guide them the data obtained and what they mean while explaining how you have presented the work or what you have plotted in the graphs. Finally, tell them what the data have revealed, what the answer to your question was. Tell them an engaging story.

Some people put a dedicated methods section between the introduction and the results, but I caution against this. Methods sections can be tedious. If they are long you can lose your audience and if there is long time before you show the data from that experiment the audience won't remember the relevant details anyway. Instead, I prefer to describe the methods and results together as discrete packets. Of course, each suggestion comes with an exception; if all your data comes from work using a single model system then you should take time to fully establish that system.

The big decision in the body part of your talk is what data to present, what order the different sections should appear, and where you should place emphasis. No surprise, the answer comes from considering what is best for your audience. When you are putting together your talk, think about how you will lead the audience through the story.

Focus on identifying what is *absolutely* required and what is more peripheral to the story, and then either get rid of the less important parts (simplifying the talk) or adjust the time you spend on each section to reflect its relative importance.

> **Big Tip**
>
> Focus your practice on the transitions between the different results slides. These need to be tight to keep your audience engaged.

When it comes to presenting your talk, the results sections usually are the parts you will find most comfortable. You will have been speaking for a little while, hopefully will have calmed down a bit and your audience should be fully invested in learning about your findings by this point. More importantly, you will know exactly what you have done and will have already thought a lot about what your results mean. When you are practising this part, I recommending focusing on making sure the transitions from one set of experiments to the next are tight to make sure that you don't lose the audience as you progress. Now is also the time to revisit the checkpoints you have identified to check that they will be delivered effectively.

I am sure you can remember a presentation when your attention drifted for a few seconds, and then you found it hard to reconnect. Therefore, the other thing to consider is whether you can provide re-entry points for anyone you lost along the way. This concept is particularly important in long talks. Of course, it would be best if you didn't lose your audience, but don't be afraid to add a slightly longer link between different sections or a mini recap to allow people re-engage. Using a slide that just states a checkpoint can also work as a re-entry point while simultaneously focusing the listeners on the critical points.

The conclusion

If you want your listeners to think that they have attended a "good" talk, then you need your conclusion to work well. Ask yourself again what aspect of your presentation your audience will value and remind them about it! You have three options here.

Concluding strategies

Option 1: Summarise

Revisit the main points you made throughout the speech. This is simple and, although it sounds a bit dull, it can be effective particularly in longer talks or when the data have taken you on a remarkable journey. Only highlight the most important points, not all the sub-points. These can simply be the checkpoints you identified at the beginning of the planning stage. If there are lots of points, you can benefit from having a summary slide and a separate conclusion slide.

A summary on its own is helpful, but to make it effective you should also connect the findings into the complete narrative in order to arrive at a wholistic conclusion.

Option 2: Where next?

Ending your talk by describing future directions can be a good option. The danger here is that these can sound like "we want to do this" rather than, "these important questions remain". The difference between these subtly different points is that the second form is an inclusive call-to-arms that makes your audience part of the solution. If you can do this well, the discussions and questions after your talk will be much more productive; you will get suggestions about how your work could progress next and benefit from the collective knowledge in the room.

Option 3: The bookend or reverse hook.

If you can bring your talk back to the hook you opened with then the audience will leave with the feeling that the story is complete. Examples could be to update the story you began with in response to the new understanding your work has added, or if you started with a fact, you could have some positive outlook on how your work could change that statistic in the future. If you posed a question in your hook or title, then you should answer that question at the end. Don't tease something then fail to deliver.

Often your conclusion can (and should) cover all these options, usually in order. However, like an introduction, a long conclusion can lose impact. Don't try to do too much; keep it short, focused and memorable. In a standard 10-12-minute conference talk, I suggest one minute as being about right, over two minutes would be too long.

Now thank the audience and open the floor for questions.

Slides and avoiding death by Powerpoint

None of the great orators in history needed slides to tell their story and, for the most part, neither do you. The reason that you might *choose* to use slides is to help deliver the more complex material, and thereby help your audience. You don't need a slide for every point. Keep this in mind as you prepare your slides.

Big Tip

Overly complicated slides do not make your story better.

Slide content rules

All writing on slides should be short. Think of your slides like tweets, stay within a 140-character limit per slide.

- Use images to replace text.
- Use graphs instead of tables.
- Use bullet points rather than full sentences.
- Limit to five words per bullet point.
- Limit to five bullet points per slide.

If someone can take everything you want to deliver from the slide without requiring any an explanation, then you have made yourself redundant, there is no need for you to be on stage at all. It's not a "talk" it's a "read"!

Do not be afraid to have a blank screen for periods; it can re-centre the audience's focus onto you and can be effective in allowing people to reconnect.

Data Slides

The main reason that you will *need* slides is to deliver data. The good news is that will have already made the content for your slides if you have kept up to date with your figure preparation. For most data sets you will be able to pick up your publication-ready figure panels and drop them into your presentation. Using the publication version will

When it comes to slide design, less is more.
Simplify your message as much as you can.
Use images and graphs instead of text or tables.
Use your slide design to focus your audience's attention.

It is ridiculous to make a slide that you then have to apologise for!

look professional and will give you the air of someone who cares about their work. This is a better way to establish credibility than telling people your history! Using your publication-ready figures will also be quick and easy, another good reason to make your figures as soon as the data collection is complete.

Unfortunately, as usual we need an exception. When you publish a paper, you know that the readers have as long as they need to go over the intricacies of the data whereas in a talk you will be more focused on delivering the *story* of the data, you are likely to want to move

Big Tip

We cannot simultaneously read and listen.
Cut out as much writing as possible.

more quickly. If your data sets are complicated, you will need to find ways to simplify them. This simplification could be something easy like cropping out some of the columns in your graphs or reducing the number of panels of images that you are showing at any one time.

This is a big point; if your slides are complicated, then your audience will either stop listening to you as they study the slide, or they will listen to you and not absorb the information. You can't read and listen at the same time.

Colour

> ## Slide Design Guideline
>
> Keep the main design features simple so that it is your message that stands out rather than the design
> - One background colour and two element colours maximum.
> - Solid colour or subtle gradient background.
> - Two font types maximum: one for headings, one for body text.
> - Maintain consistent colour/font/design throughout the deck.

If you are likely to be presenting your work in a light room, then a light background with your text and other elements in dark colours will provide maximum impact. Whereas if the room will be dark (or you intend to ask for the lights to be lowered), then a dark background with light elements will be best.

You can choose to switch from light to dark and vice versa occasionally within your slide deck. For example, if I have just a couple of slides with fluorescent microscopy images, then I might shift to a dark background for those slides and then shift back to light background for my graphs, blots etc.

Colour Combination Suggestions

Dark background:
Background: navy or dark purple.
Text/Graphics: white or yellow.
Accent colours: lime green, camel orange or light blue.

Light background:
Background: warm beige
Text/Graphics: dark blue, black, dark purple.
Accent colours: dark green, or burgundy.

Shifting between light and dark can also be used as a strategy to increase impact of a specific point or to re-centre the focus of your audience. For example, if most of your slides are light background, consider using a dark background for the key point slides.

Fonts

The best fonts for a presentation are different from those used for print. For slides, you are looking for fonts that are easy to read at a distance, with large lettering in short stretches of text. Note that the default fonts Calibri, Times and Arial are OK, but they were designed for small scale print and are not ideal for presentation. Moreover, because these three fonts are commonly used, some people might consider them to be a lazy option.

Consider trying some of the better options listed in the box to the right.

Good Fonts
Helvetica
Garamond
Futura
Gill Sans
Rockwell

Bad Fonts
Comic Sans
Papyrus

Animations

I have a mixed message with slide animations. If you use them well, then a small amount of animation can be very effective in focusing your audience's attention. But if you use animations excessively, then they can distract from your talk rather than enhancing it. As with the rest of your talk preparation, you should ask yourself "what value does this animation add to my audience enjoyment?" If the answer isn't clear, then it is better not to animate that element. Animations can also be polarising in their opinion; some academics will really dislike their use. You are unlikely to annoy anyone by not animating your slides, therefore I recommend taking a safe option and animating fewer elements when giving a job or fellowship talk.

Delivery

Handling nerves

It's not just you, everyone gets nervous before a big talk! In advance of the talk, try to focus on the positive possibilities the talk will bring rather than focusing on negatives you can't control. Picture things going well. Remember that the crowd are interested in what you have to say, they wouldn't have come otherwise.

A lot of the pre-talk nerves come from the fear of the unknown, therefore removing some of the unknowns can help you to enjoy the experience. One way to remove unknowns is to *practice.* Whenever you have a presentation coming up, make sure you plan enough time to go through your speech often enough to be sure what you will say. Indeed, very few people can give a compelling, impactful and on-time talk without practising, so even if you are unfazed by public speaking, you should still practice. The decision about whether you should script entirely and then learn your talk or whether you can have a general plan without the specifics, depends on your feel for the subject and your confidence.

While you should have a few run-throughs of the entire talk, I recommend focusing most of your practice on specific areas. Unsurprisingly, the start is vital, not only is it essential to get the audience on your side early, but also at this point the adrenaline will be pumping the hardest, so you are more likely to race through and not say everything that you planned. I suggest that you learn the first few slides by heart and practice them the most so that there is no *thinking* going on while you are trying to deliver them. Practice them in the shower, on the walk to work, anytime you have a spare minute. The conclusion slide should also get similar attention, this is simply because the conclusion is really really important! It should be perfect. There is also a danger when you get to the conclusions that you will be just be so happy to be finished that you race through it to try and get off the stage. The parts in the middle are usually less of a problem, you know your data, therefore the practice there is making sure the transitions are smooth.

Arrive early. To try and settle the nerves Remove as many of the doubts about the room or the presentation as you can. For example, loading your talk onto the computer, checking the slides look good on the screen, and that animations or videos will play smoothly will remove any technical doubts you have. In large conferences, there will usually be technical assistance available and a *speaker ready room*, where you upload your talk and check it works on their system before your session. If you arrive early to the presentation room you can also settle some nerves by identifying the stairs you will be climbing, how you will advance the slides, where the microphones are and if there is a pointer or timer available etc.

Introduce yourself to the session chair when you arrive. This will help them to know who to speak to, and it will often be them who selected your abstract for presentation, so you can know that you will have at least one person who thinks your work is interesting! Likely they'll also ask the first question if the people in the room are slow to get going.

Speak to one person (at a time). A big crowd can be intimidating so don't try to speak to them all, instead talk to individuals. Your nerves will likely be worst at the beginning, so when you first go up to the podium try to identify someone toward the middle of the room that is smiling and then deliver the first part of your talk to them. As you gain confidence, you will be able to do the same thing to more and more people. Making eye contact will let you identify how engaged the people are, which will help you identify if you should speed up or slow down. If you focus on doing this, then you will also avoid the big, and surprisingly common, problem of delivering the talk to the slides rather than the audience.

Use deep breathing during the previous talk. Taking long, deep breaths will slow your heart rate helping to calm you down. Rather than worrying out what you are going to say, you can focus on thinking about breathing. It works.

The opening – first impressions

First impressions count. Get the audience to like you, and they will be more likely to engage. Evolution has established a series of gestures that our brains register as being indicative of friend or foe. Within the first 30 seconds of your talk, your audience is subconsciously processing the signals you are giving off and will decide whether they "like" you or not. Help them come to this conclusion by looking out to the crowd and smiling. Better still, adopt an open stance, use open-handed gestures. Humanity's transition to standing on our hind legs exposes all the sensitive organs of the abdomen. Covering the belly is, therefore, a strongly ingrained defence mechanism. Opening your gestures to allow access to the belly sends evolutionary programmed signals that you are not threatened by the audience, that you are a friend.

Timing and Pace

Part of the *respect your audience* mantra is keeping to time. When you practice your talk and are working out how long to spend on each element, I recommend erring on the side of being too short rather than too long.

Big Tip

Respect your audience.

Keep to your allotted time.

If you choose to go over your allotted slot, it is a signal to your audience that you believe your time to be more important than theirs. Importantly, if you run over your allotted time in a conference setting, then the chair of the session will hurry you up or might even cut you off. If this happens, you might not get time to deliver your carefully planned concluding remarks and your audience will leave without the complete story. Disaster! If you don't have time to finish your talk, you are not just being rude you are also being ineffective and have wasted your listeners time even further! In contrast, if your talk is a little bit shorter than the allotted time, there is no real problem.

The pace of delivery is a different challenge. Too fast and too slow are equally problematic. The target is probably a little slower than a normal conversation. Most people, myself included, tend to talk too fast and, certainly, I tend to speak more slowly in practice than when giving the real thing. Being aware of your tendencies can help you adjust...the second half of my talks are often slower than the first half!

Don't be afraid of silence. If you put a striking image on the screen or ask your audience to look at a graph or other data display, you might benefit from a short stretch of silence while you know that everyone isn't listening anyway.

Make a conscious effort to direct your speech out toward
the audience rather than facing the slides.
Make eye contact.
Be extra careful if you decide to use a pointer.

Attention grabbers

Irrespective of how engaging your story is, your audience will lose focus at some point.

The data from teaching research are quite compelling; if you introduce a distinct change in your presentation roughly every ten minutes, you can effectively reset the audience focus. In a lecture, you might ask the students a question or have a discussion period to break up the time. However, that doesn't necessarily work well in a talk to a large audience.

What you are looking for is a change of focus. If you have a long stretch of data-filled slides, where the audience will be looking intently at the slides, then you could try creating a break by putting a plain black or white slide in the middle of the stretch of data. This will shift the audience's attention back to focusing entirely upon you for that period. This short break will, in turn, allow you to return to the data again with a refreshed audience. Including animations, short videos or, even just inserting an anecdote or little bit of humour can achieve the same effect. Anything that provides a little jolt to wake people up and switch them on again can help.

When you first step on stage, make an effort to
smile and use open gestures to establish a rapport
with your audience.
Don't hide behind the lectern, break down those
barriers.
Wearing a crop top to show your belly is optional!

Answering questions

Breathe. The main talk is over, now the Professor in the front row will tear you to shreds. Except, that isn't what they are trying to do. People ask questions because they are interested and because they want to know more. The questions you are asked can give you new ideas and can give you instant feedback on any part of your talk that wasn't clear. Questions are a good thing.

Sadly, this is the part of the talk where there is a limit as to how much you can practice. One option is to ask your colleagues to suggest questions that they would ask, but there are always going to be things that they haven't thought of. You are going to have to depend on your grasp of the results and your knowledge of the field. Your best preparation is reading the literature.

When you first receive a question, your mind is likely to start racing trying not only to come up with the answer but also trying to work out what the deeper implications of that question are. Rather than rushing in to answer what you *think* the real question is, try to force yourself to answer whatever the person has actually asked. You can offer the extend answer too, but only once you provide an answer to the actual question.

My advice, is that your first response should be to confirm that you have

Big Tip

Remember to thank the people who asked questions.

understood the question correctly. You can do this by repeating the question and this will also give the rest of the audience a second chance to hear what has been asked. This is particularly important in larger auditoriums, where there aren't roaming mics or when a question comes from the front of the room and spoken in a way that the people at the back won't have heard.

It is OK not to know the answer. Nobody can know everything. You should, know precisely what you did and what your data show and you should know the immediate research area very well (but once you go a bit wider, it is perfectly reasonable for your answer to start "That's not something I have considered …". Similarly, if the question asks whether you have done some specific experiment that you haven't done, the answer can simply be something like, "Not yet, but thanks for the suggestion."

There's no point giving a speech if you can't be heard. Speak loudly and clear, possibly a little slower than usual. Always use a microphone if available, they are there for a reason! Even if you are loud enough for the room you should remember that people with hearing aids might be tuned in to the closed loop system.

After the talk

Not everyone will be confident enough to ask their questions in front of the whole room, or there may simply not be enough time for all the questions. Therefore, at the end of the session, it is a good idea to hang back a little when leaving the auditorium to allow time for people to meet you. During this time, thank the organisers or chair of the session again. You always want to leave a good impression.

Bad Talk Bingo

A final comment on presentations. At this point in your career, I am sure you have been to talks or lectures that were awful for one reason or another. Let's make a pact to avoid these howlers:

Bad Talk Bingo

Loads of text	Too complex slides	Comic Sans	Laser pointer swirling
Gawdy Colours	Running over time	Reading the slides	Excessive Jargon
No Conclusions	Umms and Aaaahs	Talking too quietly	Long answer to simple question
Too fast or too slow	Skipping slides	Too small fonts	Cheesey animations

Winning feels like losing!

9.2 Posters

Remember in school when you were told to design a poster and you wondered why you needed those skills? Well, it turns out that at scientific conferences, there are not enough time slots for everybody to talk. Therefore, the organisers will arrange poster sessions to allow everyone to have an opportunity to present their work. In your career you are likely to produce many professional posters and there can be a lot of value gained by doing so effectively.

Poster sessions can be great for getting a deep, meaningful and in-depth, interaction with people who care about your work. They represent a chance to speak one-to-one with experienced people in your field. Often it is in front of a poster that new collaborations are born. Indeed, they are so useful that some conferences allow you to present a poster in addition to your talk.

Note: the instructions in this section refer primarily to presenting your research in poster format rather than just general information about making posters.

One poster with two jobs

A quick point before we begin. Your poster may have to do two jobs. First, it will be a visual aid for you when discussing your data with others, i.e. it replaces the slide deck of a presentation. Second, it might have to stand alone to present your work when you are not available to talk about it. In most conferences, you are allocated a specific time to present, but the poster will remain on display for the rest of the conference and people will interact it when you are not there.

The good news is that most of the approaches to making a good poster will be the same for both roles. However, it is worth keeping these dual roles in mind while you are deciding what you need to include.

Content and structure

A scientific poster will contain all the contents of a research paper, but these elements should be delivered primarily in pictorial form rather than using text. Your poster will need to deliver the title, an author list, a short motivation, background and rationale section, the aims and objectives or hypothesis, a truncated materials and methods section, the results (the largest and most important part), conclusions and perhaps a short amount of discussion, and any key references. The challenge is to design your poster so it can deliver the main points of your story in about five minutes, but has the details available to allow deeper discussion wherever desired. It should be attractive to the eye so that people will choose to interact with it,

To keep your story short and punchy, you will first need to decide what are the most important things you want to deliver. Once you have done that, every subsequent decision should be designed to help your reader absorb those main points. These decisions influence everything from the sizing of the elements, to selecting colour schemes,

Big Tip
You may not be able to present everything. Focus on the big picture.

fonts and to how the different parts are laid out.

Scientific Posters

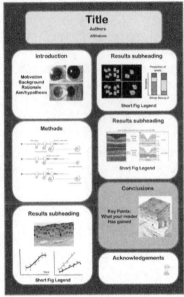

Make the title stand out so people walking past can easily read it

Use short, focused, bullet points rather than blocks of text

Use diagrams rather than text for any methods sections (if needed)

Use graphs rather than tables and include images instead of text

Use shorter, declarative titles and sub-titles rather than descriptive titles

Be consistent with font choices: choose one font and size for all headings (sans serif) and one for body text (serif).

Choose high contrast colours and remember to consider colour blind people

Conclusions are important. Consider drawing extra attention to them with an accent colour

Leave a border and space between elements so your poster doesn't feel cramped

Don't forget references, acknowledgements and funder information

Title

You need to grab people's attention. Poster titles work best if they are short, simple and definitive. You want passers-by to identify the topic area of your work, quickly and then they will come a bit closer and linger for longer if it is relevant to them. A declarative statement-type title is likely to be more memorable (as usual). Avoid using a question unless you will have the answer prominently displayed on your poster. Indeed, it is usually better to avoid having any punctuation in your title.

Make your title big and easy to read. Do not use all capitals as they take longer for a reader to absorb.

Layout

The layout of the different elements of your poster should be focused on achieving a clear flow through the narrative, to help guide those looking at it. The order in which you want the different elements to be read should be clear to the reader (I sometimes number my boxes or have arrows guiding the reader).

There are lots of ways in which posters can work, but the most obvious way is to have two (portrait) or three (landscape) columns. The reader then will progress from top to bottom down one column then down the next. Simple. Common variations on this theme are segmenting into boxes; introduction/abstract/aims and methods along the top, the results in the middle, and discussion, acknowledgements along the bottom.

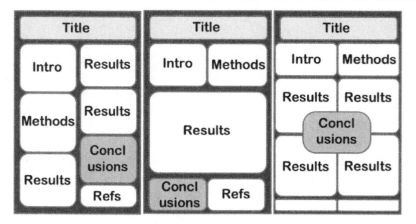

Common poster layouts.
Your layout should help the audience follow the story and should draw their eye to the most important points (usually the conclusions)

These simple options, while fine, may not be the best for *your* data. Indeed, as most posters will look something like the options above, mixing it up can help make an impact or lasting impression.

Something you should also consider is what will be in the eye-line of people walking past. If you have compelling images, you could consider having those located centrally to draw people in. You could also draw attention to specific elements by having them encased in a call-out or using a colour switch (e.g. the grey boxes for conclusions in the examples above). If you do decide to break from one of the simple traditional templates, make sure there is a clear progression through the poster, that it doesn't just feel like a jumbled mess!

Remember, whichever layout you choose, you don't use every little bit of space to deliver information. As a rough rule of thumb, about 40% of your poster should actually be empty. A poster that is crammed full will put people off It is much better to have regions of unused space to draw attention to the parts you want people to focus on.!

Replace text with images wherever you can

People are coming to your poster to see and discuss your work rather than to *read* about your work. Posters are all about delivering the message at speed rather than the minute detail (they'll get the detail from your published paper). Think about ways in which you can cut down the text on the poster

So, how can you cut down the text? For your introductions and

> **Big Tip**
> Most people don't want to read large bodies of text at a poster session. Try to find ways to replace text with pictures and diagrams.

conclusions, the obvious answer is to use short bullet points rather than full sentences or paragraphs. For your introduction; four bullet points should be all you need. #1 to establish the motivation, #2 for specific, relevant background, #3 to identify the gap in the literature, and #4 to state your aim, hypothesis or objective as appropriate.

Each bullet does not have to be a full sentence, nor should it be. The shorter they are, the better. An image to provide context or a diagram to establish anything complicated is worth including. If you can replace one or more bullet point with an image, then that is even better!

Materials and Methods sections are another area where it is quite easy to use a diagram to replace text. Otherwise, keep focused on the relevant material, use bullet points and short statements. If you are presenting data from many different approaches (and therefore would need a large block of text), an option is to split the methods and place them in figure legends. This has the advantage of providing experimental detail in the appropriate context an often works better than having dedicated methods section. People won't read figure legends unless they are interested, but, for the most part, the details of the methods are not the reason that people have come to speak to you, so it makes sense to put them in an area where they won't necessarily get the most attention.

For the Results sections, use figures all the way. If you have made the figures with full labelling, then you shouldn't need much writing. Usually, the only text you will need are for headings and figure legends. Most of the time you can just put in the individual panels you have made for your figures. However, you might need slight tweaks, for example increasing the size of the fonts for axis labels. Graphs are always better than tables, and sometimes it can be beneficial to simplify a data set to make it easier to

> **Big Tip**
>
> Most of your data figures can be the same as for publication but simplify them if it will help you to deliver your message.

deliver. Results sub-headings can be used to establish what the figure shows but, as usual, I suggest choosing a statement of your findings as a more efficient way of advancing the narrative and helping the people looking at the poster.

For the Discussion, I would again recommend short bullet points and for the Conclusion aim for a single sentence that broadly agrees or expands upon the title.

Design

Software

Posters can be easily produced in Microsoft PowerPoint or Apple Keynote. Just set the page dimension to the size you will print the poster and then you can go ahead and create your poster quickly and easily. You might even find templates made by your department are available to download which will speed up the process.

If you want to take more time and care, and create a more professional feel, then publishing programs like Adobe Illustrator, Corel Draw or Microsoft Publisher are specifically designed for this sort of task. These programmes give some added functionality that can allow you to produce more impactful content. This comes with the caveat that they might take longer to prepare at first if you are not already familiar with the software, but the end effect might be better.

Whatever you use, always do a test print before printing at poster-size and check carefully for random print marks, underlined spelling mistakes or other sorts of non-visual things that differences in background colours that are carried through to paper but which you can't see on the screen. Errors like this are embarrassing at full size.

Big Tip

Turn on gridlines. They will help you maintain spacing and align elements

Size and orientation

You will receive information about what size the poster boards will be. Make sure you check the instructions. If your poster ends up being too big it will spill over the edge and will look very unprofessional. Landscape or portrait; again, check the instructions, don't just assume it is the same as last time. Before you begin, set your slide or page dimensions so that they fit the conference requirements. This way you will be able to work at 100% scale and are able to check for pixilation of images.

Fonts

Use 2-3 fonts maximum:

- Titles and sub-titles: use a sans-serif font, e.g. **Franklin Gothic, Gill Sans, Tahoma).** Fonts without the little ticks (serifs) on the end of the letters are easier to absorb in large fonts and short phrases.
- Body text: use a serif font (e.g. Garamond, Big Caslon, Minion) fonts with serifs are easier to read in blocks of text.
- Don't use comic sans (no one will take you seriously) and avoid Calibri or other default fonts if you want your work to stand out. The default fonts are so widely used that they feel ubiquitous.

Keep font sizes consistent between the different elements.
Suggestions:

- Title: 90 pt.
- Section headings: 60 pt.
- Body text and subheadings: 30 pt.
- Figure legends and references 24 pt. (18 pt min).
- Axis and figure labels 18 pt. (final size, check after you resize).

Print the poster at A4 size and check that you can read everything at arm's length (key point).

Don't use ALL CAPITALS or underlining; these are more difficult to read, instead use **bold***ing* to draw attention to specific points where necessary.

Colours

Use one background colour and a maximum of two accent colours. Establish which accent colour is primary (titles/headings) and which is secondary and then keep consistent.

If you know where the posters will be displayed, consider selecting the background colour to work with the lighting in the poster hall. Use a light background for bright rooms, and darker backgrounds for darker places. An exception to this broad suggestion is that if most of your data are images that are predominantly black on white then consider a darker background to maximise image contrast.

Your colour scheme might be defined by your Institution or Department. Following the same branding is usually enforced when a group of people will be presenting at a large conference. There could be good news, if there is a defined colour scheme there will probably be pre-existing templates available for you.

Images

o   Make sure you are using good quality images of the highest available resolution.

o   At the final print-size, check that all images are a minimum of 150 dpi, any smaller and they may look pixelated. For example, for a 600 x 300 dpi microscopy image, the max size for print would be 10 x 5 cm.

o   Check that any graphs or tables you have inserted as tiffs are also sufficiently high resolution.

o   Remember to consider colour blind people: avoid red/green combinations.

Tables

o   It is better to use a graph wherever possible.

o   Use formatting as for publication (there is a reason why print journals use their specific format rules!).

    o   No vertical lines.

    o   Horizontal line at top and bottom.

    o   Horizontal line between column headings and data.

    o   If you think zebra stripes will help, then use very subtly different colours.

Presenting your poster

During the conference, you will have an allocated period where you should stand in front of the poster and be prepared to discuss or present it to other attendees. These poster presentations can be one-to-one or one to a crowd of people, but either way, they are much more flexible and fluid than a podium presentation. As such, you must be prepared and willing to respond to what your audience wants.

Poster sessions are often judged, and there are prizes available. The judges will be looking for good science, clear communication and for professionalism. The awards are not the reason for being there, but these judging criteria are the same elements you should aim to achieve in all presentations anyway.

It is a good idea to have a short general talk prepared (but you will always have to be flexible on the day). Aim for about five minutes or fewer to cover all the critical points in enough detail to deliver the message. There are likely to be hundreds or even thousands of posters, and people will want to look at many of those posters in the allotted time. If your stock presentation takes 30 minutes, then your audience will be unimpressed. Give them the main story then let them decide where they want to probe into in more depth.

Let your audience guide how you present your work. Before you begin presenting, ask a few small questions to gauge where to pitch your story. If it is a lab head who works in your field that is talking to you, then you won't need to spend much time establishing the background as they already know that bit, similarly they likely they will not need (or want) simple experiments explained to them in detail. However, if it is a student in a different discipline a bit more preamble is likely to be required.

As for talks, make sure the motivation, aims and conclusions are all covered in your stock presentation. People need to know why and what you wanted to achieve before you launch into discussing the minutiae.

When viewing posters, it's OK to say "sorry, I need to move on, I have other posers I would like to see"

When presenting, respect your audience by keeping your talk short and focused. Let your audience decide where they want to delve deeper

Chapter 10: Resources

10.1 Specific Materials and Methods Sections Requirments

Below are some of the examples for specific materials and methods sections that are commonly used in biology papers. These lists can be used as a checklist to ensure your methods are suitably complete.

Patient or Participant Recruitment

Details to include in your methods:

- **Where**: from which population and where were adverts distributed.
- **When**: actual date ranges.
- **How**: email, flyers, advert, targeted messages etc.
- **Inclusion and exclusion criteria**: even if there were no exclusion criteria you should explicitly state that.
- **Ethical approval**: who awarded it, what were the reference numbers for people to check.
- **How consent was obtained**: "written informed consent", parent/guardian etc.
- **How participants were separated into groups**: how was randomisation achieved, what were the demographic boundaries or other criteria used for stratification.
- **How sample sizes were decided:** details of your power analyses etc.
- **Anything else relevant to data analysis**

Note that the actual demographics of the recruited participants are usually best described in the results sections, often in table format.

Antibodies

Details to include;

- Species.
- Clonality and type (mono/polyclonal/nanobody).
- Clone name if appropriate (usually for monoclonal antibodies).
- Final concentration in the techniques ($\mu g/ml$ rather than dilution where possible).
- Where you got it from (commercial source or collaborator details).
- References for where they were generated and validated.
- Don't forget to include the details about your secondaries as well.

If the epitopes matter to the interpretation of your data, include those details too. Ask yourself (with this and with other experiments) what details would you need to interpret the data. For example, if you are looking at a transmembrane protein, knowing whether your antibodies bind inside or outside the cell could make a difference to the data interpretation. Or, if you are doing a blot and see processing products then an NH_2-terminal Ab could reveal COOH terminal processing or vice versa. Wherever the details matter, include them!

Antibodies validation data

Should you put your antibody validation data in your results, or describe them in your methods and put the data in your supplementary data? Both options can work. The answer for your work will depend on where the antibodies came from and therefore how much validation you need to describe. If they have been widely published, then your validation is probably just r positive and negative controls and those should go in the results and then you should include references to the papers with the initial validation. If the antibodies are relatively new and you have done more extensive validation e.g. performed experiments with the sole intention of determining specificity, then I would put those data in supplemental figures.

If you made any antibodies yourself then you will need to be as comprehensive as possible; epitope, carrier, boost schedule, primary screening approach, secondary screening, details of hybridoma lines etc and the validation figures.

Cell lines / Cell Culture Conditions

Required details:

General

- Media formulations and supplier's information including all supplements.
- Seeding densities for specific experimental procedures (cells per flask or dish size).
- Time from seeding to experiment.
- Plasticware supplier.
- Description of any substrate treatment: coating protein name, concentration, supplier, coating approach.
- Independent experimental unit: how many "biological repeats" how many "technical repeats" per experiment.

Immortalised/permanent lines:

- Line name, species, origin tissue, type.
- Supplier (commercial source details or collaborator).
- Reference(s) for first description/characterisation.
- How *you* validated them; STR genetic fingerprinting, blotting/ immunofluorescence/ flow, RT-qPCR? Usually the data associated with validation goes in supplementary figures.
- Passage number, feeding schedule, etc.
- Immortalisation technique (SV40T, E6/E7, hTERT etc).

Primary isolated cells

- Source tissue / supplier.
- Number of donors and donor characteristics (age, sex, disease status etc).
- Passage numbers, or even better, cell doublings.
- Ethical approval details.
- Mycoplasma screening approach (kit, primer and PCR details).
- Validation: images/blots/flow data etc. References to support these choices. See also the antibodies section for information associated with any antibodies-

based validation.

Note; for PhD thesis/dissertations: the isolation/establishment of a new primary cells or immortalised line might be integral to your data and it might be appropriate to include the characterisation details as figures in your main results rather than as supplemental material.

RNA Isolation, RT-PCR, qPCR and Endpoint PCR

The description of RNA and DNA approaches have the least flexibility of all the methods as there are standard ways for reporting quantitative PCR data that must be followed; the MIQE standards. This might sound constraining, but actually means you don't need to make any decisions, no thinking required, just follow the rules.

Even if you are not using quantitative PCR techniques, many of the MIQE guidelines still apply as good practice. The lists below are for easy reference.

Terminology

- "Reference genes" or "reference transcripts" are preferred rather than housekeeping genes for your calibrators.
- qPCR and RT-qPCR are used for quantitative approaches (do not use RT to mean "real-time", RT stands for reverse transcriptase).
- Quantification is a word. quantitation is not!
- C_q is the preferred term for the threshold point of detection. Do not use any of the other versions used by different light cycler manufacturers.

Experimental design

Follow the usual rules, define:

- The experimental groups and control groups.
- What you have considered to be the independent experimental unit.
- The number of biological replicates (the sample size).
- The number of technical replicates.

Each of the following subsections can be written as a separate paragraph following the order below. Dealing with one step at a time can help ensure you cover all the needed details.

Sample

Describe all the steps taken in going from cells/tissue to RNA/DNA:

- Description of source material, mass of tissue, dissection steps, also include anything else that could be relevant such as post-mortem time.
- Describe time in storage before processing and how stored. If frozen; how and how quickly. If fixed, how, which buffers etc.
- Describe processing procedures: name of kit or reagents and any modifications.
- Details of DNase or RNAse treatments.
- Describe how you determined nucleic acid quantification (instrument and method). Include yields.
- How you assessed purity e.g. A260/280, and the thresholds you accepted for

continuing.

- RNA integrity method/instrument: electrophoresis traces, RIN of 3'/5' transcripts and associated thresholds for continuing.

Reverse Transcription

Describe the complete reaction conditions:

- Amount of RNA and reaction volume.
- Oligonucleotide used for priming (oligo dT/ random hexamers etc).
- Reverse transcriptase and concentration.
- Temperature and time.
- Storage conditions of cDNA.

qPCR target information and Oligonucleotides

- Gene symbol and accession number.
- Location of amplicon and amplicon length.
- Include details on which splice isoforms are amplified.
- In silico specificity screen (BLAST etc).
- Primer sequences (and probe sequences if used), references if required.
- Location and identity of any modifications.
- Purification method.

If you have multiple oligonucleotides, this is a situation where it can be helpful to use a table! Make sure all the details go into the table.

qPCR protocol

- Complete reaction conditions.
- Reaction volume + amount of cDNA or DNA.
- Concentrations of primers, $MgCl_2$, dNTP, probes, + other additives, polymerase identity and concentration, buffer/kit identity and manufacturer + chemical compositions.
- Information about plates, tubes, etc.
- Complete thermocycling parameters.

Every time you have used a kit for an experiment you should be able to explain what is in that kit and what is happening at each stage. Kits are great, but you must understand your science.

Assay Validation

- Method for C_q determination.
- Specificity; gel, sequence, melt curve*, digest.
- Results for no template controls.
- Calibration curve with slope and y-intercept: PCR efficiency calculated from slope + r^2*.
- Outlier identification and disposition.
- Justification of number and choice of reference transcripts.
- Number and concordance of biological replicates.
- Number and stage of technical replicates (i.e. reverse transcription or qPCR

repeats).

- Repeatability and reproducibility.
- Power analysis.
- Statistical methods for significance determination, software.
- C_q or raw data submission.

melt curves and efficiency curves are critical for evaluating your data. Many students have told me that they did not generate these data as the primers were validated by a company or were published elsewhere. That is not good enough to publish and you should be setting yourself higher standards. Your priming efficiency depends not just on the sequence but also on the cDNA quality and integrity, upon the machine you are using and its cycling parameters and on presence of PCR inhibitors in your sample. Melt and efficiency curves should accompany every publication in the supplementary figures.

Immunocytochemistry or immunohistochemistry

Sample preparation

Immunohistochemistry (tissue staining):

- Where the samples came from and how it was acquired (include ethics and patient recruitment section for human samples); age, disease/genotype etc.
- Time from dissection to processing and/or post-mortem time to retrieval.
- How samples were dissected, oriented etc.
- Post-acquisition processing: fixation, sucrose infiltration, dehydration procedures.
- Embedding: paraffin. OCT etc
- Sectioning: thickness, microtome/cryostat used, slide type.
- Rehydration.
- Antigen retrieval – time, chemical or enzyme, (pH if relevant), how (microwave, water bath etc).

Immunocytochemistry (cell staining):

- All the usual cell culture requirements (usually in their own section).
- Time from plating, density, and substrate; glass/plastic, thickness, coating etc.
- Fixation: chemical, time, temperature.
- Extraction: chemical, time, temperature.
- Storage if relevant.

Processing

- Block: substance (e.g. goat serum), concentration, dilution, time and temperature of incubation.
- Primary antibodies incubation: all the antibodies details (often in their own subsection see above), dilution and solution, incubation time, temperature.
- Wash steps: solution, concentration, time.
- Secondary antibodies incubations: all the antibodies details, dilution, solution, incubation time and temperature.
- Equivalent details for stains: concentrations, incubations, chemicals etc.
- Mounting and coverslipping: solutions, counterstains etc.

- Don't forget to describe your controls: make sure you include full details here (and supplemental figures for review).
  - Non-specific binding controls: pre-immune serum, isotypes, antigen cleared).
  - Negative controls: different species, non-expressing cells-knockdown, knockout.
  - Positive controls: known cell lines/tissue with established distribution. Include the images in your manuscript if you are trying to prove a negative.

Imaging

- Microscope information: manufacturer, objective, numerical aperture of objectives, immersion, illumination.
- Acquisition details: laser line details, filters, Z stack thickness, pinhole size, binning, scan direction, pixel, anything else relevant to analysis.
- Camera details: manufacturer, colour profile, bit depth

Analysis

Do not forget this bit!

- How did you go from images to numbers? What did you quantify?
- If you didn't quantify anything, how did you choose which images to use?
- Which programme did you use, including plug-ins etc. for FIJI (Image J)?
- If you performed any sort of intensity analysis, then described what you used to calibrate; what were your internal controls?
- If using an indirect or enzymatic approach, how did you account for non-stoichiometric binding?
- Which method of co-localisation quantification you used (Mander's, Pearson's etc.), you can't say signals are co-distributed without providing numerical data to support the statement.
- Answer the usual statistical questions: number and types of replicates, technical, biological repeats performed.
- If you used pseudo-colouring or look up tables (LUTs), then describe the algorithm (remember, it is best to show black and white images in black and white, and don't forget to consider how colour-blind people will view your figures.)

Writing about immunoblotting

How western blots / immunoblots methods and data are reported has come under a lot of scrutiny recently and new guides to best practice are being implemented in all the reputable journals. This means that the standards reviewers/examiners will expect is higher than has previously been published. This is a good thing, it means that the papers you read now will be better, but it also means you must work to these higher standards if you want to be taken seriously and for your data to stand up to scrutiny.

As always, it is the details that matter. Here are things you should always report:

- **Sample preparation**: buffer, volumes/mass or volume/area or volume/cell, times to extraction, etc must be fully detailed.

- **Running, gel and transfer conditions:** gel type and percentages, buffers, transfer type and duration, membrane type, pore size and brand. Native/reduced.
- **Blocking:** solution used, time.
- **Antibodies:** primary, secondary, catalogue numbers, supplier, lot number, concentration.
- **Incubation:** times and temperatures for both antibody steps and in what buffers.
- **Washing:** solution, time, how often.
- **Detection:** reagents and imaging techniques, scan times/laser intensities etc including brand of X-ray film.
- **Molecular mass of band of interest**: which markers did you use (indicate marker location on blot). Most good journals now require the original uncropped blot to be uploaded with the manuscript.
- **Balancing loading;** If you quantify immunoblots then you must clearly explain how protein loading was normalised between lanes.
  - Normalisation to total protein loading is preferred but which total protein stain did you use?
  - If you used a reference protein for quantification or to balance loading, you need to provide the evidence that that reference protein does not change in your experimental conditions.
- **Quantification:**
  - Your methods should detail how the linear ranges of your antibodies were determined (include as supplemental figure).
  - Signals obtained using antibodies to phosphorylated epitopes should be normalised to total protein level of the target protein
  - Describe the software you used
  - How! Compared what to what, plotted as what?
- **Replicates**: technical and biological repeats numbers and how you defined independence (i.e. what you count as a biological repeat) must be defined.
- **Statistics:** as always, which tests, what comparisons, what threshold did you decide to accept as significant?

10.2 Experimental Design Checklists
Correlative Studies

Main research question / Aim:

Project hypothesis:

Experimental
research objective:

What key comparisons
will be made?

What experimental system
will be used?
☐ Human ☐ Animal model ☐ 3D culture ☐ Primary culture ☐ Cell line ☐ In vitro ☐ Other

Do you have the necessary
ethical approvals?
☐ Yes ☐ No ☐ Not required

What manipulation(s)
will you introduce? (if any)

Is manipulation possible
in this system? ☐ Yes ☐ No
Is this manipulation
biological meaningful? ☐ Yes ☐ No

What will be the
experimental unit?

Are these truly independent? ☐ Yes ☐ No

What measurements
will be made?

Are these? ☐ Direct ☐ Indirect

What confounding variables or
alternative interpretations do
you need to control for?

Can you control for: Reverse causation? ☐ Yes ☐ No Third variables? ☐ Yes ☐ No

What controls do you need?
☐ Experimental positive control
☐ Experimental negative control
☐ Biological positive control
☐ Biological negative control
☐ Calibration control
☐ Technical repeats

How will you confirm
identified correlations?

How many independent
variables do you have? [   ] Are these ☐ Continuous ☐ Discrete ☐ Binary

How many outcome
variables do you have? [   ] Do you have a primary
outcome variable? ☐ Yes ☐ No

What type of data will your
outcome variable(s) give? ☐ Continuous ☐ Discrete ☐ Binary How many times will
You measure each sample? ☐ Once ☐ Repeated

What statistical test do
you expect to use?

How strong a correlation would
be biological meaningful? [   ] How much variability
do you expect? [   ]

What p value will you consider significant (type I error / false
discovery rate)? ☐ 0.1 % ☐ 1 % ☐ 5 % ☐ Other

What power do you require? (type II error rate)? ☐ 70 % ☐ 80 % ☐ 90 % ☐ Other

What experimental N do you require? [   ] Is that N
possible? ☐ Yes ☐ No

How long will it take? [   ] How much will it cost? [   ] Are these
numbers feasible? ☐ Yes ☐ No

How will you confirm that
identified correlations are
meaningful? (follow up)

Manipulative Studies

Main research question / Aim:

Project hypothesis:

Experimental research question:

Experimental hypothesis:

Is your hypothesis testable? ☐ Yes ☐ No Is your hypothesis specific? ☐ Yes ☐ No

What key comparisons will be made?

What experimental system will be used?
☐ Human ☐ Animal model ☐ 3D culture ☐ Primary culture ☐ Cell line ☐ In vitro ☐ Other

What manipulation(s) will you introduce?

Is manipulation possible in this system? ☐ Yes ☐ No Is this manipulation biological meaningful? ☐ Yes ☐ No

Do you have the necessary ethical approvals? ☐ Yes ☐ No ☐ Not required

---

What will be the experimental unit?

Are these truly independent? ☐ Yes ☐ No

What measurements will be made?

Are these? ☐ Direct ☐ Indirect

What confounding variables or alternative interpretations do you need to control for?

What controls do you need?
☐ Experimental positive control
☐ Experimental negative control
☐ Biological positive control
☐ Biological negative control
☐ Calibration control
☐ Technical repeats

How many independent variables do you have? ____ Are these ☐ Continuous ☐ Discrete ☐ Binary

How many outcome variables do you have? ____ Do you have a primary outcome variable? ☐ Yes ☐ No

What type of data will your outcome variable(s) give? ☐ Continuous ☐ Discrete ☐ Binary How many times will You measure each sample? ☐ Once ☐ Repeated

What statistical test do you expect to use?

How big a difference would be biological meaningful? (effect size) ____ How much variability between groups do you expect? ____

What p value will you consider significant (type I error rate)? ☐ 0.1% ☐ 1% ☐ 5% ☐ Other

What power do you require (type II error rate)? ☐ 70% ☐ 80% ☐ 90% ☐ Other

---

What experimental N do you require (use an online calculator)? ____ Is that N possible? ☐ Yes ☐ No

How long will it take? ____ How much will it cost? ____ Are these numbers feasible? ☐ Yes ☐ No

10.3 Statistics Tests Flow Chart

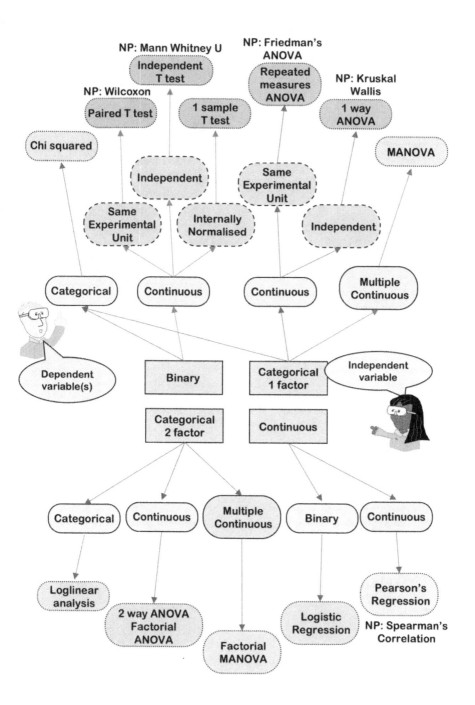

10.4 Choosing A Graph

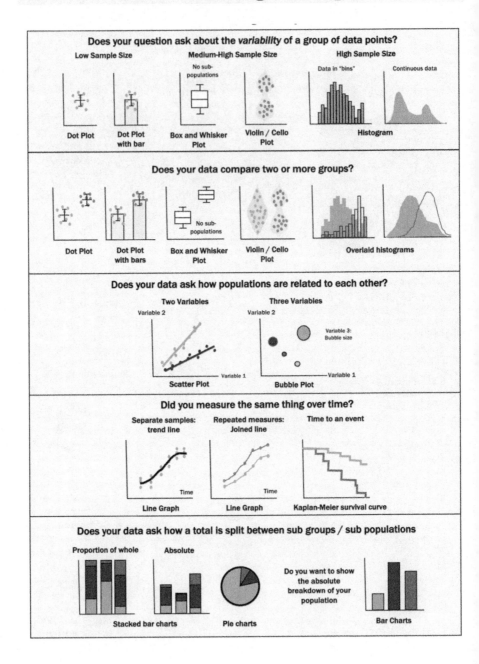

10.5 Manuscript Preparation Checklist

Abstract

- ☐ Motivation
- ☐ Aim/hypothesis
- ☐ Methods
- ☐ Results
- ☐ Conclusions
- ☐ Within word count

Introduction

- ☐ Title keyword In first sentence
- ☐ 1st para: study motivation explicit
- ☐ States research question clearly
- ☐ Sets up discussion
- ☐ Fully referenced
- ☐ Topic sentences tell the story

Materials and Methods

- ☐ Past tense throughout
- ☐ Written in prose
- ☐ All techniques fully described
- ☐ Supplier information described
- ☐ Data analysis described
- ☐ Journal specific requirements addressed
- ☐ Fully referenced
- ☐ Stats methods described
- ☐ SI units compliance

Results In each subsection

- ☐ Experiment goals explicit
- ☐ Methods sentence included
- ☐ Subsection headings as result statements (where possible)
- ☐ Results described
- ☐ Numerical data included
- ☐ Fully referenced
- ☐ All figure panels referred to
- ☐ Figures referred to in order 1A, 1B, 2A, 2B

Discussion

- ☐ Short synopsis
- ☐ Comments on accuracy of hypothesis
- ☐ Fully referenced
- ☐ Major finding receives most space
- ☐ Implications discussed
- ☐ Limitations acknowledged
- ☐ Conclusions paragraph

10.6 Figure Preparation Checklist

Figures

Supplemental

	1	2	3	4	5	6	7	1	2	3	4
Journal requirements followed	☐	☐	☐	☐	☐	☐	☐	☐	☐	☐	☐
Font sizes (2 max) [＿＿＿pt]	☐	☐	☐	☐	☐	☐	☐	☐	☐	☐	☐
Fonts: Arial, Times, symbol	☐	☐	☐	☐	☐	☐	☐	☐	☐	☐	☐
Lines: widths used [＿＿＿pt]	☐	☐	☐	☐	☐	☐	☐	☐	☐	☐	☐
Graphs: Y axis labels accurate	☐	☐	☐	☐	☐	☐	☐	☐	☐	☐	☐
Graphs: Y axis units indicated	☐	☐	☐	☐	☐	☐	☐	☐	☐	☐	☐
Graphs: X axis labels complete	☐	☐	☐	☐	☐	☐	☐	☐	☐	☐	☐
Graphs: titles	☐	☐	☐	☐	☐	☐	☐	☐	☐	☐	☐
Colour scheme: max contrast	☐	☐	☐	☐	☐	☐	☐	☐	☐	☐	☐
Colour scheme: suitable for colour blind	☐	☐	☐	☐	☐	☐	☐	☐	☐	☐	☐
Elements sized appropriately	☐	☐	☐	☐	☐	☐	☐	☐	☐	☐	☐
Printed and checked	☐	☐	☐	☐	☐	☐	☐	☐	☐	☐	☐
Understandable without need for legend	☐	☐	☐	☐	☐	☐	☐	☐	☐	☐	☐

Figure Legends

	1	2	3	4	5	6	7	1	2	3	4
Title clause (bold)	☐	☐	☐	☐	☐	☐	☐	☐	☐	☐	☐
All panels described	☐	☐	☐	☐	☐	☐	☐	☐	☐	☐	☐
Methods clear	☐	☐	☐	☐	☐	☐	☐	☐	☐	☐	☐
Description of graphs	☐	☐	☐	☐	☐	☐	☐	☐	☐	☐	☐
Graphs: Y axis units indicated	☐	☐	☐	☐	☐	☐	☐	☐	☐	☐	☐
Graphs: X axis labels complete	☐	☐	☐	☐	☐	☐	☐	☐	☐	☐	☐
Description of experimental N	☐	☐	☐	☐	☐	☐	☐	☐	☐	☐	☐
Stats tests indicated	☐	☐	☐	☐	☐	☐	☐	☐	☐	☐	☐
Scale bars defined	☐	☐	☐	☐	☐	☐	☐	☐	☐	☐	☐
Acronyms and labels defined	☐	☐	☐	☐	☐	☐	☐	☐	☐	☐	☐

Printed in Great Britain
by Amazon